The Report on Unidentified Flying Objects

Project Blue Book - The Complete 1956 Report on UFOs by an Officer of the U.S. Air Force

By Edward J. Ruppelt

PANTIANOS
CLASSICS

Published by Pantianos Classics

ISBN-13: 978-1-78987-233-0

First published in 1956

Contents

Foreword

This is a book about unidentified flying objects—UFO's—"flying saucers." It is actually more than a book; it is a report because it is the first time that anyone, either military or civilian, has brought together in one document all the facts about this fascinating subject. With the exception of the style, this report is written exactly the way I would have written it had I been officially asked to do so while I was chief of the Air Force's project for investigating UFO reports—Project Blue Book.

In many instances I have left out the names of the people who reported seeing UFO's, or the names of certain people who were associated with the project, just as I would have done in an official report. For the same reason I have changed the locale in which some of the UFO sightings occurred. This is especially true in chapter fifteen, the story of how some of our atomic scientists detected radiation whenever UFO's were reported near their "UFO-detection stations." This policy of not identifying the "source," to borrow a term from military intelligence, is insisted on by the Air Force so that the people who have co-operated with them will not get any unwanted publicity. Names are considered to be "classified information."

But the greatest care has been taken to make sure that the omission of names and changes in locale has in no way altered the basic facts because this report is based on the facts—all of the facts—nothing of significance has been left out.

It was only after considerable deliberation that I put this report together, because it had to be told accurately, with no holds barred. I finally decided to do it for two reasons. First, there is world- wide interest in flying saucers; people want to know the facts. But more often than not these facts have been obscured by secrecy and confusion, a situation that has led to wild speculation on one end of the scale and an almost dangerously blas? attitude on the other. It is only when all of the facts are laid out that a correct evaluation can be made.

Second, after spending two years investigating and analyzing UFO reports, after talking to the people who have seen UFO's— industrialists, pilots, engineers, generals, and just the plain man- on-the-street, and after discussing the subject with many very capable scientists, I felt that I was in a position to be able to put together the complete account of the Air Force's struggle with the flying saucer.

The report has been difficult to write because it involves something that doesn't officially exist. It is well known that ever since the first flying saucer was reported in June 1947 the Air Force has officially said that there is no proof that such a thing as an interplanetary spaceship exists. But what is not well known is that this conclusion is far from being unanimous among the military and their scientific advisers because of the one word, *proof*; so the UFO investigations continue.

The hassle over the word "proof" boils down to one question: What constitutes proof? Does a UFO have to land at the River Entrance to the Pentagon, near the Joint Chiefs of Staff offices? Or is it proof when a ground radar station detects a UFO, sends a jet to intercept it, the jet pilot sees it, and locks on with his radar, only to have the UFO streak away at a phenomenal speed? Is it proof when a jet pilot fires at a UFO and sticks to his story even under the threat of court-martial? Does this constitute proof?

The at times hotly debated answer to this question may be the answer to the question, "Do the UFO's really exist?"

I'll give you the facts—all of the facts—you decide.

July 1955,
E. J. RUPPELT

Chapter One - Project Blue Book and the UFO Story

In the summer of 1952 a United States Air Force F-86 jet interceptor shot at a flying saucer.

This fact, like so many others that make up the full flying saucer story, has never before been told.

I know the full story about flying saucers and I know that it has never before been told because I organized and was chief of the Air Force's Project Blue Book, the special project set up to investigate and analyze unidentified flying object, or UFO, reports. (UFO is the official term that I created to replace the words "flying saucers.")

There is a fighter base in the United States which I used to visit frequently because, during 1951, 1952, and 1953, it got more than its share of good UFO reports.

The commanding officer of the fighter group, a full colonel and command pilot, believed that UFO's were real. The colonel believed in UFO's because he had a lot of faith in his pilots—and they had chased UFO's in their F-86's. He had seen UFO's on the scopes of his radar sets, and he knew radar.

The colonel's intelligence officer, a captain, didn't exactly believe that UFO's were real, but he did think that they warranted careful investigation. The logic the intelligence officer used in investigating UFO reports—and in getting answers to many of them— made me wish many times that he worked for me on Project Blue Book.

One day the intelligence officer called me at my base in Dayton, Ohio. He wanted to know if I was planning to make a trip his way soon. When I told him I expected to be in his area in about a week, he asked me to be sure to look him up. There was no special hurry, he added, but he had something very interesting to show me.

When we got wind of a good story, Project Blue Book liked to start working on it at once, so I asked the intelligence officer to tell me what he had. But nothing doing. He didn't want to discuss it over the phone. He even vetoed the idea of putting it into a secret wire. Such extreme caution really stopped me, because anything can be coded and put in a wire.

When I left Dayton about a week later I decided to go straight to the fighter base, planning to arrive there in midmorning. But while I was changing airlines my reservations got fouled up, and I was faced with waiting until evening to get to the base. I called the intelligence officer and told him about the mix-up. He told me to hang on right there and he would fly over and pick me up in a T-33 jet.

As soon as we were in the air, on the return trip, I called the intelligence officer on the interphone and asked him what was going on. What did he have? Why all the mystery? He tried to tell me, but the interphone wasn't working too well and I couldn't understand what he was saying. Finally he told me to wait until we returned to his office and I could read the report myself.

Report! If he had a UFO report why hadn't he sent it in to Project Blue Book as he usually did?

We landed at the fighter base, checked in our parachutes, Mae Wests, and helmets, and drove over to his office. There were several other people in the office, and they greeted me with the usual question, "What's new on the flying saucer front?" I talked with them for a while, but was getting impatient to find out what was on the intelligence officer's mind. I was just about to ask him about the mysterious report when he took me to one side and quietly asked me not to mention it until everybody had gone.

Once we were alone, the intelligence officer shut the door, went over to his safe, and dug out a big, thick report. It was the standard Air Force reporting form that is used for all intelligence reports, including UFO reports. The intelligence officer told me that this was the only existing copy. He said that he had been told to destroy all copies, but had saved one for me to read.

With great curiosity, I took the report and started to read. What *had* happened at this fighter base?

About ten o'clock in the morning, one day a few weeks before, a radar near the base had picked up an unidentified target. It was an odd target in that it came in very fast—about 700 miles per hour— and then slowed down to about 100 miles per hour. The radar showed that it was located northeast of the airfield, over a sparsely settled area.

Unfortunately the radar station didn't have any height-finding equipment. The operators knew the direction of the target and its distance from the station but they didn't know its altitude. They reported the target, and two F-86's were scrambled.

The radar picked up the F-86's soon after they were airborne, and had begun to direct them into the target when the target started to fade on the radarscope. At the time several of the operators thought that this fade was caused by the target's losing altitude rapidly and getting below the radar's beam. Some of the other operators thought that it was a high-flying target and that it was fading just because it was so high.

In the debate which followed, the proponents of the high-flying theory won out, and the F-86's were told to go up to 40,000 feet. But before the aircraft could get to that altitude, the target had been completely lost on the radarscope.

The F-86's continued to search the area at 40,000 feet, but could see nothing. After a few minutes the aircraft ground controller called the F-86's and told one to come down to 20,000 feet, the other to 5,000 feet, and continue the search. The two jets made a quick letdown, with one pilot stopping at 20,000 feet and the other heading for the deck.

7

The second pilot, who was going down to 5,000 feet, was just beginning to pull out when he noticed a flash below and ahead of him. He flattened out his dive a little and headed toward the spot where he had seen the light. As he closed on the spot he suddenly noticed what he first thought was a weather balloon. A few seconds later he realized that it couldn't be a balloon because it was staying ahead of him. Quite an achievement for a balloon, since he had built up a lot of speed in his dive and now was flying almost straight and level at 3,000 feet and was traveling "at the Mach."

Again the pilot pushed the nose of the F-86 down and started after the object. He closed fairly fast, until he came to within an estimated 1,000 yards. Now he could get a good look at the object. Although it had looked like a balloon from above, a closer view showed that it was definitely round and flat—saucer-shaped. The pilot described it as being "like a doughnut without a hole."

As his rate of closure began to drop off, the pilot knew that the object was picking up speed. But he pulled in behind it and started to follow. Now he was right on the deck.

About this time the pilot began to get a little worried. What should he do? He tried to call his buddy, who was flying above him somewhere in the area at 20,000 feet. He called two or three times but could get no answer. Next he tried to call the ground controller but he was too low for his radio to carry that far. Once more he tried his buddy at 20,000 feet, but again no luck.

By now he had been following the object for about two minutes and during this time had closed the gap between them to approximately 500 yards. But this was only momentary. Suddenly the object began to pull away, slowly at first, then faster. The pilot, realizing that he couldn't catch *it*, wondered what to do next.

When the object traveled out about 1,000 yards, the pilot suddenly made up his mind—he did the only thing that he could do to stop the UFO. It was like a David about to do battle with a Goliath, but he had to take a chance. Quickly charging his guns, he started shooting. . . . A moment later the object pulled up into a climb and in a few seconds it was gone. The pilot climbed to 10,000 feet, called the other F-86, and now was able to contact his buddy. They joined up and went back to their base.

As soon as he had landed and parked, the F-86 pilot went into operations to tell his story to his squadron commander. The mere fact that he had fired his guns was enough to require a detailed report, as a matter of routine. But the circumstances under which the guns actually were fired created a major disturbance at the fighter base that day.

After the squadron commander had heard his pilot's story, he called the group commander, the colonel, and the intelligence officer. They heard the pilot's story.

For some obscure reason there was a "personality clash," the intelligence officer's term, between the pilot and the squadron commander. This was obvious, according to the report I was reading, because the squadron commander immediately began to tear the story apart and accuse the pilot of "cracking up," or of just "shooting his guns for the hell of it and using the wild story as a cover-up."

Other pilots in the squadron, friends of the accused pilot— including the intelligence officer and a flight surgeon—were called in to "testify." All of these men were aware of the fact that in certain instances a pilot can "flip" for no good reason, but none of them said that he had noticed any symptoms of mental crack-up in the unhappy pilot.

None, except the squadron commander. He kept pounding home his idea— that the pilot was "psycho"—and used a few examples of what the report called "minor incidents" to justify his stand.

Finally the pilot who had been flying with the "accused" man was called in. He said that he had been monitoring the tactical radio channel but that he hadn't heard any calls from his buddy's low- flying F-86. The squadron commander triumphantly jumped on this point, but the accused pilot tended to refute it by admitting he was so jumpy that he might not have been on the right channel. But when he was asked if he had checked or changed channels after he had lost the object and before he had finally contacted the other F-86, he couldn't remember.

So ended the pilot's story and his interrogation.

The intelligence officer wrote up his report of a UFO sighting, but at the last minute, just before sending it, he was told to hold it back. He was a little unhappy about this turn of events, so he went in to see why the group commander had decided to delay sending the report to Project Blue Book.

They talked over the possible reactions to the report. If it went out it would cause a lot of excitement, maybe unnecessarily. Yet, if the pilot actually had seen what he claimed, it was vitally important to get the report in to ATIC immediately. The group commander said that he would made his decision after a talk with his executive officer. They decided not to send the report and ordered it destroyed.

When I finished reading, the intelligence officer's first comment was, "What do you think?"

Since the evaluation of the report seemed to hinge upon conflicts between personalities I didn't know, I could venture no opinion, except that the incident made up the most fascinating UFO report I'd ever seen. So I batted the intelligence officer's question back to him.

"I know the people involved," he replied, "and I don't think the pilot was nuts. I can't give you the report, because Colonel ——— told me to destroy it. But I did think you should know about it." Later he burned the report.

The problems involved in this report are typical. There are certain definite facts that can be gleaned from it; the pilot did see something and he did shoot at something, but no matter how thoroughly you investigate the incident that something can never be positively identified. It might have been a hallucination or it might have been some vehicle from outer space; no one will ever know. It was a UFO.

9

The UFO story started soon after June 24, 1947, when newspapers all over the United States carried the first flying saucer report. The story told how nine very bright, disk-shaped objects were seen by Kenneth Arnold, a Boise, Idaho, businessman, while he was flying his private plane near Mount Rainier, in the state of Washington. With journalistic license, reporters converted Arnold's description of the individual motion of each of the objects—like "a saucer skipping across water"—into "flying saucer," a name for the objects themselves. In the eight years that have passed since Arnold's memorable sighting, the term has become so common that it is now in Webster's Dictionary and is known today in most languages in the world.

For a while after the Arnold sighting the term "flying saucer" was used to describe all disk-shaped objects that were seen flashing through the sky at fantastic speeds. Before long, reports were made of objects other than disks, and these were also called flying saucers. Today the words are popularly applied to anything seen in the sky that cannot be identified as a common, everyday object.

Thus a flying saucer can be a formation of lights, a single light, a sphere, or any other shape; and it can be any color. Performance-wise, flying saucers can hover, go fast or slow, go high or low, turn 90- degree corners, or disappear almost instantaneously.

Obviously the term "flying saucer" is misleading when applied to objects of every conceivable shape and performance. For this reason the military prefers the more general, if less colorful, name: unidentified flying objects. UFO (pronounced Yoo-foe) for short.

Officially the military uses the term "flying saucer" on only two occasions. First in an explanatory sense, as when briefing people who are unacquainted with the term "UFO": "UFO—you know—flying saucers." And second in a derogatory sense, for purposes of ridicule, as when it is observed, "He says he saw a flying saucer."

This second form of usage is the exclusive property of those persons who positively know that all UFO's are nonsense. Fortunately, for the sake of good manners if for no other reason, the ranks of this knowing category are constantly dwindling. One by one these people drop out, starting with the instant they see their first UFO.

Some weeks after the first UFO was seen on June 24, 1947, the Air Force established a project to investigate and analyze all UFO reports. The attitude toward this task varied from a state of near panic, early in the life of the project, to that of complete contempt for anyone who even mentioned the words "flying saucer."

This contemptuous attitude toward "flying saucer nuts" prevailed from mid-1949 to mid-1950. During that interval many of the people who were, or had been, associated with the project believed that the public was suffering from "war nerves."

Early in 1950 the project, for all practical purposes, was closed out; at least it rated only minimum effort. Those in power now reasoned that if you didn't mention the words "flying saucers" the people would forget them and the saucers would go away. But this reasoning was false, for instead of vanishing, the UFO reports got better and better.

Airline pilots, military pilots, generals, scientists, and dozens of other people were reporting UFO's, and in greater detail than in reports of the past. Radars, which were being built for air defense, began to pick up some very unusual targets, thus lending technical corroboration to the unsubstantiated claims of human observers.

As a result of the continuing accumulation of more impressive UFO reports, official interest stirred. Early in 1951 verbal orders came down from Major General Charles P. Cabell, then Director of Intelligence for Headquarters, U.S. Air Force, to make a study reviewing the UFO situation for Air Force Headquarters.

I had been back in the Air Force about six months when this happened. During the second world war I had been a B-29 bombardier and radar operator. I went to India, China, and later to the Pacific, with the original B-29 wing. I flew two DCF's, and some Air Medals' worth of missions, got out of the Air Force after the war, and went back to college. To keep my reserve status while I was in school, I flew as a navigator in an Air Force Reserve Troop Carrier Wing.

Not long after I received my degree in aeronautical engineering, the Korean War started, and I went back on active duty. I was assigned to the Air Technical Intelligence Center at Wright-Patterson Air Force Base, in Dayton, Ohio. ATIC is responsible for keeping track of all foreign aircraft and guided missiles. ATIC also had the UFO project.

I had just finished organizing a new intelligence group when General Cabell's order to review past UFO reports came down. Lieutenant Colonel Rosengarten, who received the order at ATIC, called me in and wanted to know if I'd take the job of making the review. I accepted.

When the review was finished, I went to the Pentagon and presented my findings to Major General Samford, who had replaced General Cabell as Director of Intelligence.

ATIC soon got the word to set up a completely new project for the investigation and analysis of UFO reports. Since I had made the review of past UFO reports I was the expert, and I got the new job. It was given the code name Project Blue Book, and I was in charge of it until late in 1953. During this time members of my staff and I traveled close to half a million miles. We investigated dozens of UFO reports, and read and analyzed several thousand more. These included every report ever received by the Air Force.

For the size of the task involved Project Blue Book was always understaffed, even though I did have ten people on my regular staff plus many paid consultants representing every field of science. All of us on Project Blue Book had Top Secret security clearances so that security was no block in our investigations. Behind this organization was a reporting network made up of every Air Force base intelligence officer and every Air Force radar station in the world, and the Air Defense Command's Ground Observer Corps. This reporting net sent Project Blue Book reports on every conceivable type of UFO, by every conceivable type of person.

What did these people actually see when they reported that they had observed a UFO? Putting aside truly unidentifiable flying objects for the present, this question has several answers.

In many instances it has been positively proved that people have reported balloons, airplanes, stars, and many other common objects as UFO's. The people who make such reports don't recognize these common objects because something in their surroundings temporarily assumes an unfamiliar appearance.

Unusual lighting conditions are a common cause of such illusions. A balloon will glow like a "ball of fire" just at sunset. Or an airplane that is not visible to the naked eye suddenly starts to reflect the sun's rays and appears to be a "silver ball." Pilots in F- 94 jet interceptors chase Venus in the daytime and fight with balloons at night, and people in Los Angeles see weird lights.

On October 8, 1954, many Los Angeles newspapers and newscasters carried an item about a group of flying saucers, bright lights, flying in a V formation. The lights had been seen from many locations over Southern California. Pilots saw them while bringing their airplanes into Los Angeles International Airport, Air Force pilots flying out of Long Beach saw them, two CBS reporters in Hollywood gave an eyewitness account, and countless people called police and civil defense officials. All of them excitedly reported lights they could not identify. The next day the Air Force identified the UFO's; they were Air Force airplanes, KC-97 aerial tankers, refueling B-47 jet bombers in flight. The reason for the weird effect that startled so many Southern Californians was that when the refueling is taking place a floodlight on the bottom of the tanker airplane lights up the bomber that is being refueled. The airplanes were flying high, and slowly, so no sound was heard; only the bright floodlights could be seen. Since most people, even other pilots, have never seen a night aerial refueling operation and could not identify the odd lights they saw, the lights became UFO's.

In other instances common everyday objects look like UFO's because of some odd quirk in the human mind. A star or planet that has been in the sky every day of the observer's life suddenly "takes off at high speed on a highly erratic flight path." Or a vapor trail from a high-flying jet—seen a hundred times before by the observer—becomes a flying saucer.

Some psychologists explain such aberrations as being akin to the crowd behavior mechanism at work in the "bobby-sox craze." Teen-agers don't know why they squeal and swoon when their current fetish sways and croons. Yet everybody else is squealing, so they squeal too. Maybe that great comedian, Jimmy Durante, has the answer: "Everybody wants to get into the act." I am convinced that a certain percentage of UFO reports come from people who see flying saucers because others report seeing them.

But this "will to see" may have deeper roots, almost religious implications, for some people. Consciously or unconsciously, they want UFO's to be real and to come from outer space. These individuals, frightened perhaps by threats of atomic destruction, or lesser fears—who knows what—act as if nothing that men can do can save the earth. Instead, they seek salvation from outer space, on the forlorn premise that flying saucer men, by their very existence, are wiser and more advanced than we. Such people may reason that a race of men capable of interplanetary travel have lived well into, or through, an atomic age. They have survived and they can tell us their secret of survival. Maybe the threat of an atomic war unified their planet and allowed them to divert their war effort to one of social and technical advancement. To such people a searchlight on a cloud or a bright star is an interplanetary spaceship.

If all the UFO reports that the Air Force has received in the past eight years could be put in this "psychological quirk" category, Project Blue Book would never have been organized. It is another class of reports that causes the Air Force to remain interested in UFO's. This class of reports are called "Unknowns."

In determining the identity of a UFO, the project based its method of operation on a well-known psychological premise. This premise is that to get a reaction from one of the senses there must be a stimulus. If you think you see a UFO you must have seen something. Pure hallucinations are extremely rare.

For anything flying in the air the stimulus could be anything that is normally seen in the air. Balloons, airplanes, and astronomical bodies are the commoner stimuli. Birds and insects are common also, but usually are seen at such close range that they are nearly always recognized. Infrequently observed things, such as sundogs, mirages, huge fireballs, and a host of other unusual flying objects, are also known stimuli.

On Project Blue Book our problem was to identify these stimuli. We had methods for checking the location, at any time, of every balloon launched anywhere in the United States. To a certain degree the same was true for airplanes. The UFO observer's estimate of where the object was located in the sky helped us to identify astronomical bodies. Huge files of UFO characteristics, along with up-to-the- minute weather data, and advice from specialists, permitted us to identify such things as sun-dogs, paper caught in updrafts, huge meteors, etc.

This determination of the stimuli that triggered UFO sightings, while not an insurmountable task, was a long, tedious process. The identification of known objects was routine, and caused no excitement. The excitement and serious interest occurred when we received UFO reports in which the observer was reliable and the stimuli could not be identified. These were the reports that challenged the project and caused me to spend hours briefing top U.S. officials. These were the reports that we called "Unknowns."

Of the several thousand UFO reports that the Air Force has received since 1947, some 15 to 20 per cent fall into this category called unknown. This

13

means that the observer was not affected by any determinable psychological quirks and that after exhaustive investigation the object that was reported could not be identified. To be classed as an unknown, a UFO report also had to be "good," meaning that it had to come from a competent observer and had to contain a reasonable amount of data.

Reports are often seen in the newspapers that say: "Mrs. Henry Jones, of 5464 South Elm, said that 10:00A.M. she was shaking her dust mop out of the bedroom window when she saw a flying saucer"; or "Henry Armstrong was driving between Grundy Center and Rienbeck last night when he saw a light. Henry thinks it was a flying saucer." This is not a good UFO report.

This type of UFO report, if it was received by Project Blue Book, was stamped "Insufficient Data for Evaluation" and dropped into the dead file, where it became a mere statistic.

Next to the "Insufficient Data" file was a file marked "C.P." This meant crackpot. Into this file went all reports from people who had talked with flying saucer crews, who had inspected flying saucers that had landed in the United States, who had ridden in flying saucers, or who were members of flying saucer crews. By Project Blue Book standards, these were not "good" UFO reports either.

But here is a "good" UFO report with an "unknown" conclusion:

On July 24, 1952, two Air Force colonels, flying a B-25, took off from Hamilton Air Force Base, near San Francisco, for Colorado Springs, Colorado. The day was clear, not a cloud in the sky.

The colonels had crossed the Sierra Nevada between Sacramento and Reno and were flying east at 11,000 feet on "Green 3," the aerial highway to Salt Lake City. At 3:40P.M. they were over the Carson Sink area of Nevada, when one of the colonels noticed three objects ahead of them and a little to their right. The objects looked like three F-86's flying a tight V formation. If they were F-86's they should have been lower, according to civil air regulations, but on a clear day some pilots don't watch their altitude too closely.

In a matter of seconds the three aircraft were close enough to the B-25 to be clearly seen. They were not F-86's. They were three bright silver, delta wing craft with no tails and no pilot's canopies. The only thing that broke the sharply defined, clean upper surface of the triangular wing was a definite ridge that ran from the nose to the tail.

In another second the three deltas made a slight left bank and shot by the B-25 at terrific speed. The colonels estimated that the speed was at least three times that of an F-86. They got a good look at the three deltas as the unusual craft passed within 400 to 800 yards of the B-25.

When they landed at Colorado Springs, the two colonels called the intelligence people at Air Defense Command Headquarters to make a UFO report. The suggestion was offered that they might have seen three F-86's. The colonels promptly replied that if the objects had been F-86's they would have easily been recognized as such. The colonels knew what F-86's looked like.

Air Defense Command relayed the report to Project Blue Book. An investigation was started at once.

Flight Service, which clears all military aircraft flights, was contacted and asked about the location of aircraft near the Carson Sink area at 3:40P.M. They had no record of the presence of aircraft in that area.

Since the colonels had mentioned delta wing aircraft, and both the Air Force and the Navy had a few of this type, we double-checked. The Navy's deltas were all on the east coast, at least all of the silver ones were. A few deltas painted the traditional navy blue were on the west coast, but not near Carson Sink. The Air Force's one delta was temporarily grounded.

Since balloons once in a while can appear to have an odd shape, all balloon flights were checked for both standard weather balloons and the big 100-foot-diameter research balloons. Nothing was found.

A quick check on the two colonels revealed that both of them were command pilots and that each had several thousand hours of flying time. They were stationed at the Pentagon. Their highly classified assignments were such that they would be in a position to recognize *anything* that the United States knows to be flying anywhere in the world.

Both men had friends who had "seen flying saucers" at some time, but both had openly voiced their skepticism. Now, from what the colonels said when they were interviewed after landing at Colorado Springs, they had changed their opinions.

Nobody knows what the two colonels saw over Carson Sink. However, it is always possible to speculate. Maybe they just thought they were close enough to the three objects to see them plainly. The objects might have been three F-86's: maybe Flight Service lost the records. It could be that the three F-86's had taken off to fly in the local area of their base but had decided to do some illegal sight-seeing. Flight Service would have no record of a flight like this. Maybe both of the colonels had hallucinations.

There is a certain mathematical probability that any one of the above speculative answers is correct—correct for this one case. If you try this type of speculation on hundreds of sightings with "unknown" answers, the probability that the speculative answers are correct rapidly approaches zero.

Maybe the colonels actually did see what they thought they did, a type of craft completely foreign to them.

Another good UFO report provides an incident in which there is hardly room for any speculation of this type. The conclusion is more simply, "Unknown," period.

On January 20, 1952, at seven-twenty in the evening, two master sergeants, both intelligence specialists, were walking down a street on the Fairchild Air Force Base, close to Spokane, Washington.

Suddenly both men noticed a large, bluish-white, spherical-shaped object approaching from the east. They stopped and watched the object carefully, because several of these UFO's had been reported by pilots from the air base over the past few months. The sergeants had written up the reports on these earlier sightings.

The object was traveling at a moderately fast speed on a horizontal path. As it passed to the north of their position and disappeared in the west, the sergeants noted that it had a long blue tail. At no time did they hear any sound. They noted certain landmarks that the object had crossed and estimated the time taken in passing these landmarks. The next day they went out and measured the angles between these landmarks in order to include them in their report.

When we got the report at ATIC, our first reaction was that the master sergeants had seen a large meteor. From the evidence I had written off, as meteors, all previous similar UFO reports from this air base.

The sergeants' report, however, contained one bit of information that completely changed the previous picture. At the time of the sighting there had been a solid 6,000-foot-thick overcast at 4,700 feet. And meteors don't go that low.

A few quick calculations gave a rather fantastic answer. If the object was just at the base of the clouds it would have been 10,000 feet from the two observers and traveling 1,400 miles per hour.

But regardless of the speed, the story was still fantastic. The object was no jet airplane because there was no sound. It was not a searchlight because there were none on the air base. It was not an automobile spotlight because a spotlight will not produce the type of light the sergeants described. As a double check, however, both men were questioned on this point. They stated firmly that they had seen hundreds of searchlights and spotlights playing on clouds, and that this was not what they saw.

Beyond these limited possibilities the sergeants' UFO discourages fruitful speculation. The object remains unidentified.

The UFO reports made by the two colonels and the two master sergeants are typical of hundreds of other good UFO reports which carry the verdict, "Conclusion unknown."

Some of these UFO reports have been publicized, but many have not. Very little information pertaining to UFO's was withheld from the press—if the press knew of the occurrence of specific sightings. Our policy on releasing information was to answer only direct questions from the press. If the press didn't know about a given UFO incident, they naturally couldn't ask questions about it. Consequently such stories were never released. In other instances, when the particulars of a UFO sighting were released, they were only the bare facts about what was reported. Any additional information that might have been developed during later investigations and analyses was not released.

There is a great deal of interest in UFO's and the interest shows no signs of diminishing. Since the first flying saucer skipped across the sky in the summer of 1947, thousands of words on this subject have appeared in every newspaper and most magazines in the United States. During a six-month period in 1952 alone 148 of the nation's leading newspapers carried a total of over 16,000 items about flying saucers.

During July 1952 reports of flying saucers sighted over Washington, D.C., cheated the Democratic National Convention out of precious headline space.

The subject of flying saucers, which has generated more unscientific behavior than any other topic of modern times, has been debated at the meetings of professional scientific societies, causing scientific tempers to flare where unemotional objectivity is supposed to reign supreme.

Yet these thousands of written words and millions of spoken words— all attesting to the general interest—have generated more heat than light. Out of this avalanche of print and talk, the full, factual, true story of UFO's has emerged only on rare occasions. The general public, for its interest in UFO's, has been paid off in misinformation.

Many civilian groups must have sensed this, for while I was chief of Project Blue Book I had dozens of requests to speak on the subject of UFO's. These civilian requests had to be turned down because of security regulations.

I did give many official briefings, however, behind closed doors, to certain groups associated with the government—all of them upon request.

The subject of UFO's was added to a regular series of intelligence briefings given to students at the Air Force's Command and Staff School, and to classes at the Air Force's Intelligence School.

I gave briefings to the technical staff at the Atomic Energy Commission's Los Alamos laboratory, where the first atomic bomb was built. The theater where this briefing took place wouldn't hold all of the people who tried to get in, so the briefing was recorded and replayed many times. The same thing happened at AEC's Sandia Base, near Albuquerque.

Many groups in the Pentagon and the Office of Naval Research requested UFO briefings. Civilian groups, made up of some of the nation's top scientists and industrialists, and formed to study special military problems, worked in a UFO briefing. Top Air Force commanders were given periodic briefings.

Every briefing I gave was followed by a discussion that lasted anywhere from one to four hours.

In addition to these, Project Blue Book published a classified monthly report on UFO activity. Requests to be put on distribution for this report were so numerous that the distribution had to be restricted to major Air Force Command Headquarters.

This interest was not caused by any revolutionary information that was revealed in the briefings or reports. It stemmed only from a desire to get the facts about an interesting subject.

Many aspects of the UFO problem were covered in these official briefings. I would give details of many of the better reports we received, our conclusions about them, and how those conclusions were reached. If we had identified a UFO, the audience was told how the identification was made. If we concluded that the answer to a UFO sighting was "Unknown," the audience learned why we were convinced it was unknown.

Among the better sightings that were described fully to interested government groups were: the complete story of the Lubbock Lights, including the possible sighting of the same V-shaped light formations at other locations on the same night; the story of a group of scientists who detected mysterious nuclear radiation when UFO's were sighted; and all of the facts behind such famous cases as the Mantell Incident, the Florida scoutmaster who was burned by a "flying saucer," and headline-capturing sightings at Washington, D.C.

I showed them what few photographs we had, the majority of which everyone has seen, since they have been widely published in magazines and newspapers. Our collection of photographs was always a disappointment as far as positive proof was concerned because, in a sense, if you've seen one you've seen them all. We had no clear pictures of a saucer, just an assortment of blurs, blotches, and streaks of light.

The briefings included a description of how Project Blue Book operated and a survey of the results of the many studies that were made of the mass of UFO data we had collected. Also covered were our interviews with a dozen North American astronomers, the story of the unexplained green fireballs of New Mexico, and an account of how a committee of six distinguished United States scientists spent many hours attempting to answer the question, "Are the UFO's from outer space?"

Unfortunately the general public was never able to hear these briefings. For a long time, contrary to present thinking in military circles, I have believed that the public also is entitled to know the details of what was covered in these briefings (less, of course, the few items pertaining to radar that were classified "Secret," and the names of certain people). But withholding these will not alter the facts in any way.

A lot has already been written on the subject of UFO's, but none of it presents the true, complete story. Previous forays into the UFO field have been based on inadequate information and have been warped to fit the personal biases of the individual writers. Well meaning though these authors may be, the degree to which their books have misinformed the public is incalculable.

It is high time that we let the people know.

The following chapters present the true and complete UFO story, based on what I learned about UFO's while I was chief of Project Blue Book, the Air Force's project for the investigation and analysis of UFO reports. Here is the same information that I gave to Secretary of the Air Force, Thomas K. Finletter, to the Air Force commanders, to scientists and industrialists. This is what the Air Force knows about unidentified flying objects.

You may not agree with some of the official ideas or conclusions— neither did a lot of people I briefed—but this is the story.

Chapter Two - The Era of Confusion Begins

On September 23, 1947, the chief of the Air Technical Intelligence Center, one of the Air Force's most highly specialized intelligence units, sent a letter to the Commanding General of the then Army Air Forces.

The letter was in answer to the Commanding General's verbal request to make a preliminary study of the reports of unidentified flying objects. The letter said that after a preliminary study of UFO reports, ATIC concluded that, to quote from the letter, "the reported phenomena were real." The letter strongly urged that a permanent project be established at ATIC to investigate and analyze future UFO reports. It requested a priority for the project, a registered code name, and an over-all security classification. ATICs request was granted and Project Sign, the forerunner of Project Grudge and Project Blue Book, was launched. It was given a 2A priority, 1A being the highest priority an Air Force project could have. With this the Air Force dipped into the most prolonged and widespread controversy it has ever, or may ever, encounter. The Air Force grabbed the proverbial bear by the tail and to this day it hasn't been able to let loose.

The letter to the Commanding General of the Army Air Forces from the chief of ATIC had used the word "phenomena." History has shown that this was not a too well-chosen word. But on September 23, 1947, when the letter was written, ATICs intelligence specialists were confident that within a few months or a year they would have the answer to the question, "What are UFO's?" The question, "Do UFO's exist?" was never mentioned. The only problem that confronted the people at ATIC was, "Were the UFO's of Russian or interplanetary origin?" Either case called for a serious, secrecy-shrouded project. Only top people at ATIC were assigned to Project Sign.

Although a formal project for UFO investigation wasn't set up until September 1947, the Air Force had been vitally interested in UFO reports ever since June 24, 1947, the day Kenneth Arnold made the original UFO report.

As Arnold's story of what he saw that day has been handed down by the bards of saucerism, the true facts have been warped, twisted, and changed. Even some points in Arnold's own account of his sighting as published in his book, *The Coming of the Saucers*, do not jibe with what the official files say he told the Air Force in 1947. Since this incident was the original UFO sighting, I used to get many inquiries about it from the press and at briefings. To get the true and accurate story of what did happen to Kenneth Arnold on June 24, 1947, I had to go back through old newspaper files, official reports, and talk to people who had worked on Project Sign. By cross-checking these data and talking to people who had heard Arnold tell about his UFO sighting soon after it happened, I finally came up with what I believe is the accurate story.

Arnold had taken off from Chehalis, Washington, intending to fly to Yakima, Washington. About 3:00 P.M. he arrived in the vicinity of Mount Rainier. There was a Marine Corps C-46 transport plane lost in the Mount Rainier area, so Arnold decided to fly around awhile and look for it. He was looking down at the ground when suddenly he noticed a series of bright flashes off to his left. He looked for the source of the flashes and saw a string of nine very bright disk- shaped objects, which he estimated to be 45 to 50 feet in length. They were traveling from north to south across the nose of his airplane. They were flying in a reversed echelon (i.e., lead object high with the rest stepped down), and as they flew along they weaved in and out between the mountain peaks, once passing behind one of the peaks. Each individual object had a skipping motion described by Arnold as a "saucer skipping across water."

During the time that the objects were in sight, Arnold had clocked their speed. He had marked his position and their position on the map and again noted the time. When he landed he sketched in the flight path that the objects had flown and computed their speed, almost 1,700 miles per hour. He estimated that they had been 20 to 25 miles away and had traveled 47 miles in 102 seconds.

I found that there was a lot of speculation on this report. Two factions at ATIC had joined up behind two lines of reasoning. One side said that Arnold had seen plain, everyday jet airplanes flying in formation. This side's argument was based on the physical limitations of the human eye, visual acuity, the eye's ability to see a small, distant object. Tests, they showed, had proved that a person with normal vision can't "see" an object that subtends an angle of less than 0.2 second of arc. For example, a basketball can't be seen at a distance of several miles but if you move the basketball closer and closer, at some point you will be able to see it. At this point the angle between the top and bottom of the ball and your eye will be about 0.2 of a second of arc. This was applied to Arnold's sighting. The "Arnold-saw-airplanes" faction maintained that since Arnold said that the objects were 45 to 50 feet long they would have had to be much closer than he had estimated or he couldn't even have seen them at all. Since they were much closer than he estimated, Arnold's timed speed was all wrong and instead of going 1,700 miles per hour the objects were traveling at a speed closer to 400 miles per hour, the speed of a jet. There was no reason to believe they weren't jets. The jets appeared to have a skipping motion because Arnold had looked at them through layers of warm and cold air, like heat waves coming from a hot pavement that cause an object to shimmer.

The other side didn't buy this idea at all. They based their argument on the fact that Arnold knew where the objects were when he timed them.

After all, he was an old mountain pilot and was as familiar with the area around the Cascade Mountains as he was with his own living room. To cinch this point the fact that the objects had passed *behind* a mountain peak was brought up. This positively established the distance the objects were from Arnold and confirmed his calculated 1,700-miles-per-hour speed. Besides, no airplane can weave in and out between mountain peaks in the short time that Arnold was watching them. The visual acuity factor only strengthened the "Arnold-saw-a- flying-saucer" faction's theory that what he'd seen was a spaceship. If he could see the objects 20 to 25 miles away, they must have been about 210 feet long instead of the poorly estimated 45 to 50 feet.

In 1947 this was a fantastic story, but now it is just another UFO report marked "Unknown." It is typical in that if the facts are accurate, if Arnold actually did see the UFO's go *behind* a mountain peak, and if he knew his exact position at the time, the UFO problem cannot be lightly sloughed off; but there are always "ifs" in UFO reports. This is the type of report that led Major General John A. Samford, Director of Intelligence for Headquarters, Air Force, to make the following comment during a press conference in July 1952: "However, there have remained a percentage of this total [of all UFO reports received by the Air Force], about 20 per cent of the reports, that have come from credible observers of relatively incredible things. We keep on being concerned about them."

In warping, twisting, and changing the Arnold incident, the writers of saucer lore haven't been content to confine themselves to the incident itself; they have dragged in the crashed Marine Corps' C-46. They intimate that the same flying saucers that Arnold saw shot down the C-46, grabbed up the bodies of the passengers and crew, and now have them pickled at the University of Venus Medical School. As proof they apply the same illogical reasoning that they apply to most everything. The military never released photos of the bodies of the dead men, therefore there were no bodies. There were photographs and there were bodies. In consideration of the families of air crewmen and passengers, photos of air crashes showing dead bodies are never released.

Arnold himself seems to be the reason for a lot of the excitement that heralded flying saucers. Stories of odd incidents that occur in this world are continually being reported by newspapers, but never on the scale of the first UFO report. Occasional stories of the "Himalayan snowmen," or the "Malayan monsters," rate only a few inches or a column on the back pages of newspapers. Arnold's story, if it didn't make the headlines, at least made the front page. I had the reason for this explained to me one day when I was investigating a series of UFO reports in California in the spring of 1952.

I was making my headquarters at an air base where a fighter-bomber wing was stationed. Through a mutual friend I met one of the fighter- bomber pilots who had known Arnold. In civilian life the pilot was a newspaper reporter and had worked on the original Arnold story. He told me that when the story first broke all the newspaper editors in the area were thoroughly convinced that the incident was a hoax, and that they intended to write the story as such. The more they dug into the facts, however, and into Arnold's reputation, the more it appeared that he was telling the truth. Besides having an unquestionable character, he was an excellent mountain pilot, and mountain pilots are a breed of men who know every nook and cranny of the mountains

21

in their area. The most fantastic part of Arnold's story had been the 1,700-miles-per-hour speed computed from Arnold's timing the objects between two landmarks. "When Arnold told us how he computed the speed," my chance acquaintance told me, "we all put a lot of faith in his story." He went on to say that when the editors found out that they were wrong about the hoax, they did a complete about-face, and were very much impressed by the story. This enthusiasm spread, and since the Air Force so quickly denied ownership of the objects, all of the facts built up into a story so unique that papers all over the world gave it front-page space.

There was an old theory that maybe Arnold had seen wind whipping snow along the mountain ridges, so I asked about this. I got a flat "Impossible." My expert on the early Arnold era said, "I've lived in the Pacific Northwest many years and have flown in the area for hundreds of hours. It's impossible to get powder snow low in the mountains in June. Personally, I believe Arnold saw some kind of aircraft and they weren't from this earth." He went on to tell me about two other very similar sightings that had happened the day after Arnold saw the nine disks. He knew the people who made these sightings and said that they weren't the kind to go off "half cocked." He offered to get a T-6 and fly me up to Boise to talk to them since they had never made a report to the military, but I had to return to Dayton so I declined.

Within a few days of Arnold's sighting, others began to come in. On June 28 an Air Force pilot in an F-51 was flying near Lake Mead, Nevada, when he saw a formation of five or six circular objects off his right wing. This was about three-fifteen in the afternoon.

That night at nine-twenty, four Air Force officers, two pilots, and two intelligence officers from Maxwell AFB in Montgomery, Alabama, saw a bright light traveling across the sky. It was first seen just above the horizon, and as it traveled toward the observers it "zigzagged," with bursts of high speed. When it was directly overhead it made a sharp 90-degree turn and was lost from view as it traveled south.

Other reports came in. In Milwaukee a lady saw ten go over her house "like blue blazes," heading south. A school bus driver in Clarion, Iowa, saw an object streak across the sky. In a few seconds twelve more followed the first one. White Sands Proving Ground in New Mexico chalked up the first of the many sightings that this location would produce when several people riding in an automobile saw a pulsating light travel from horizon to horizon in thirty seconds. A Chicago housewife saw one "with legs."

The week of July 4, 1947, set a record for reports that was not broken until 1952. The center of activity was the Portland, Oregon, area. At 11:00A.M. a carload of people driving near Redmond saw four disk-shaped objects streaking past Mount Jefferson. At 1:05P.M. a policeman was in the parking lot behind the Portland City Police Headquarters when he noticed some pigeons suddenly began to flutter around as if they were scared. He looked up and saw five large disk-shaped objects, two going south and three going east. They were traveling at a high rate of speed and seemed to be oscillating about their lateral axis. Minutes later two other policemen, both ex- pilots, reported three of the same things flying in trail. Before long the harbor patrol called into headquarters. A crew of four patrolmen had seen three to six of the disks, "shaped like chrome hub caps," traveling very fast. They also oscillated as they flew. Then the citizens of Portland began to see them. A man saw one going east and two going north. At four-thirty a woman called in and had just seen one that looked like "a new dime flipping around." Another man reported two, one going southeast, one northeast. From Milwaukie, Oregon, three were reported going northwest. In Vancouver, Washington, sheriff's deputies saw twenty to thirty.

The first photo was taken on July 4 in Seattle. After much publicity it turned out to be a weather balloon.

That night a United Airlines crew flying near Emmett, Idaho, saw five. The pilot's report read:

Five "somethings," which were thin and smooth on the bottom and rough-appearing on top, were seen silhouetted against the sunset shortly after the plane took off from Boise at 8:04 P.M. We saw them clearly. We followed them in a northeasterly direction for about 45 miles. They finally disappeared. We were unable to tell whether they outsped us or disintegrated. We can't say whether they were "smearlike," oval, or anything else but whatever they were they were not aircraft, clouds or smoke.

Civilians did not have a corner on the market. On July 6 a staff sergeant in Birmingham, Alabama, saw several "dim, glowing lights" speeding across the sky and photographed one of them. Also on the sixth the crew of an Air Force B-25 saw a bright, disk-shaped object "low at nine o'clock." This is one of the few reports of an object lower than the aircraft. At Fairfield-Suisun AFB in California a pilot saw something travel three quarters of the way across the sky in a few seconds. It, too, was oscillating on its lateral axis.

According to the old hands at ATIC, the first sighting that really made the Air Force take a deep interest in UFO's occurred on July 8 at Muroc Air Base (now Edwards AFB), the supersecret Air Force test center in the Mojave Desert of California. At 10:10 A.M. a test pilot was running up the engine of the then new XP-84 in preparation for a test flight. He happened to look up and to the north he saw what first appeared to be a weather balloon traveling in a westerly direction. After watching it a few seconds, he changed his mind. He had been briefed on the high-altitude winds, and the object he saw was going against the wind. Had it been the size of a normal aircraft, the test pilot estimated that it would have been at 10,000 to 12,000 feet and traveling 200 to 225 miles per hour. He described the object as being spherically shaped and yellowish white in color.

Ten minutes before this several other officers and airmen had seen three objects. They were similar except they had more of a silver color. They were also heading in a westerly direction.

Two hours later a crew of technicians on Rogers Dry Lake, adjacent to Muroc Air Base, observed another UFO. Their report went as follows:

On the 8 July 1947 at 11:50 we were sitting in an observation truck located in Area #3, Rogers Dry Lake. We were gazing upward toward a formation of two P-82's and an A-26 aircraft flying at 20,000 feet. They were preparing to carry out a seat-ejection experiment. We observed a round object, white aluminum color, which at first resembled a parachute canopy. Our first impression was that a premature ejection of the seat and dummy had occurred but this was not the case. The object was lower than 20,000 feet, and was falling at three times the rate observed for the test parachute, which ejected thirty seconds after we first saw the object. As the object fell it drifted slightly north of due west against the prevailing wind. The speed, horizontal motion, could not be determined, but it appeared to be slower than the maximum velocity F-80 aircraft.

As this object descended through a low enough level to permit observation of its lateral silhouette, it presented a distinct oval- shaped outline, with two projections on the upper surface which might have been thick fins or nobs. These crossed each other at intervals, suggesting either rotation or oscillation of slow type.

No smoke, flames, propeller arcs, engine noise, or other plausible or visible means of propulsion were noted. The color was silver, resembling an aluminum-painted fabric, and did not appear as dense as a parachute canopy.

When the object dropped to a level such that it came into line of vision of the mountain tops, it was lost to the vision of the observers.

It is estimated that the object was in sight about 90 seconds. Of the five people sitting in the observation truck, four observed this object.

The following is our opinion about this object:

It was man-made, as evidenced by the outline and functional appearance.

Seeing this was not a hallucination or other fancies of sense.

Exactly four hours later the pilot of an F-51 was flying at 20,000 feet about 40 miles south of Muroc Air Base when he sighted a "flat object of a light-reflecting nature." He reported that it had no vertical fin or wings. When he first saw it, the object was above him and he tried to climb up to it, but his F-51 would not climb high enough. All air bases in the area were contacted but they had no aircraft in the area.

By the end of July 1947 the UFO security lid was down tight. The few members of the press who did inquire about what the Air Force was doing got the same treatment that you would get today if you inquired about the number of thermonuclear weapons stock-piled in the U.S.'s atomic arsenal. No one, outside of a few high-ranking officers in the Pentagon, knew what the people in the barbed-wire enclosed Quonset huts that housed the Air Technical Intelligence Center were thinking or doing.

The memos and correspondence that Project Blue Book inherited from the old UFO projects told the story of the early flying saucer era. These memos and pieces of correspondence showed that the UFO situation was considered to be serious; in fact, very serious. The paper work of that period also indicated the confusion that surrounded the investigation; confusion almost to the point of panic. The brass wanted an answer, quickly, and people were taking off in all directions. Everyone's theory was as good as the next and each person with any weight at ATIC was plugging and investigating his own theory. The ideas as to the origin of the UFO's fell into two main categories, earthly and non-earthly. In the earthly category the Russians led, with the U.S. Navy and their XF-5-U-1, the "Flying Flapjack," pulling a not too close second. The desire to cover all leads was graphically pointed up to be a personal handwritten note I found in a file. It was from ATIC's chief to a civilian intelligence specialist. It said, "Are you positive that the Navy junked the XF-5- U-1 project?" The non-earthly category ran the gamut of theories, with space animals trailing interplanetary craft about the same distance the Navy was behind the Russians.

This confused speculating lasted only a few weeks. Then the investigation narrowed down to the Soviets and took off on a much more methodical course of action.

When World War II ended, the Germans had several radical types of aircraft and guided missiles under development. The majority of these projects were in the most preliminary stages but they were the only known craft that could even approach the performance of the objects reported by UFO observers. Like the Allies, after World War II the Soviets had obtained complete sets of data on the latest German developments. This, coupled with rumors that the Soviets were frantically developing the German ideas, caused no small degree of alarm. As more UFO's were observed near the Air Force's Muroc Test Center, the Army's White Sands Proving Ground, and atomic bomb plants, ATIC's efforts became more concentrated.

Wires were sent to intelligence agents in Germany requesting that they find out exactly how much progress had been made on the various German projects.

The last possibility, of course, was that the Soviets had discovered some completely new aerodynamic concept that would give saucer performance.

While ATIC technical analysts were scouring the United States for data on the German projects and the intelligence agents in Germany were seeking out the data they had been asked for, UFO reports continued to flood the country. The Pacific Northwest still led with the most sightings, but every state in the Union was reporting a few flying saucers.

At first there was no co-ordinated effort to collect data on the UFO reports. Leads would come from radio reports or newspaper items. Military intelligence agencies outside of ATIC were hesitant to investigate on their own initiative because, as is so typical of the military, they lacked specific orders. When no orders were forthcoming, they took this to mean that the military had no interest in the UFO's. But before long this placid attitude changed, and changed drastically. Classified orders came down to investigate *all* UFO sightings. Get every detail and send it direct to ATIC at Wright Field. The order carried no explanation as to why the information was wanted. This lack of an explanation and the fact that the information was to be sent directly to a high-powered intelligence group within Air Force Headquarters stirred the imagination of every potential cloak-and-dagger man in the military intelligence system. Intelligence people in the field who had previously been free with opinions now clammed up tight.

The era of confusion was progressing.

Early statements to the press, which shaped the opinion of the public, didn't reduce the confusion factor. While ATIC was grimly expending maximum effort in a serious study, "certain high-placed officials" were officially chuckling at the mention of UFO's.

In July 1947 an International News Service wire story quoted the public relations officer at Wright Field as saying, "So far we haven't found anything to confirm that saucers exist. We don't think they are guided missiles." He went on to say, "As things are now, they appear to be either a phenomenon or a figment of somebody's imagination."

A few weeks later a lieutenant colonel who was Assistant to the Chief of Staff of the Fourth Air Force was widely quoted as saying, "There is no basis for belief in flying saucers in the Tacoma area [referring to a UFO sighting in the area of Tacoma, Washington], or any other area."

The "experts," in their stories of saucer lore, have said that these brush-offs of the UFO sightings were intentional smoke screens to cover the facts by adding confusion. This is not true; it was merely a lack of coordination. But had the Air Force tried to throw up a screen of confusion, they couldn't have done a better job.

When the lieutenant colonel from the Fourth Air Force made his widely publicized denunciation of saucer believers he specifically mentioned a UFO report from the Tacoma, Washington, area.

The report of the investigation of this incident, the Maury Island Mystery, was one of the most detailed reports of the early UFO era. The report that we had in our files had been pieced together by Air Force Intelligence and other agencies because the two intelligence officers who started the investigation couldn't finish it. They were dead.

For the Air Force the story started on July 31, 1947, when Lieutenant Frank Brown, an intelligence agent at Hamilton AFB, California, received a long-distance phone call. The caller was a man whom 111 call Simpson, who had met Brown when Brown investigated an earlier UFO sighting, and he had a hot lead on another UFO incident. He had just talked to two Tacoma Harbor patrolmen. One of them had seen six UFO's hover over his patrol boat and spew out chunks of odd metal. Simpson had some of the pieces of the metal.

The story sounded good to Lieutenant Brown, so he reported it to his chief. His chief OK'd a trip and within an hour Lieutenant Brown and Captain Davidson were flying to Tacoma in an Air Force B-25. When they arrived they met Simpson and an airline pilot friend of his in Simpson's hotel room. After the usual round of introductions Simpson told Brown and Davidson that he had received a letter from a Chicago publisher asking him, Simpson, to investigate this case. The publisher had paid him $200 and wanted an exclusive on the story, but things were getting too hot, Simpson wanted the military to take over.

Simpson went on to say that he had heard about the experience off Maury Island but that he wanted Brown and Davidson to hear it firsthand. He had called the two harbor patrolmen and they were on their way to the hotel. They arrived and they told their story.

I'll call these two men Jackson and Richards although these aren't their real names. In June 1947, Jackson said, his crew, his son, and the son's dog were on his patrol boat patrolling near Maury Island, an island in Puget Sound, about 3 miles from Tacoma. It was a gray day, with a solid cloud deck down at about 2,500 feet. Suddenly everyone on the boat noticed six "doughnut-shaped" objects, just under the clouds, headed toward the boat. They came closer and closer, and when they were about 500 feet over the boat they stopped. One of the doughnut-shaped objects seemed to be in trouble as the other five were hovering around it. They were close, and everybody got a good look. The UFO's were about 100 feet in diameter, with the "hole in the doughnut" being about 25 feet in diameter. They were a silver color and made absolutely no noise. Each object had large portholes around the edge.

As the five UFO's circled the sixth, Jackson recalled, one of them came in and appeared to make contact with the disabled craft. The two objects maintained contact for a few minutes, then began to separate. While this was going on, Jackson was taking photos. Just as they began to separate, there was a dull "thud" and the next second the UFO began to spew out sheets of very light metal from the hole in the center. As these were fluttering to the water, the UFO began to throw out a harder, rocklike material. Some of it landed on the beach of Maury Island. Jackson took his crew and headed toward the beach of Maury Island, but not before the boat was damaged, his son's arm had been injured, and the dog killed. As they reached the island they looked up and saw that the UFO's were leaving the area at high speed. The harbor patrolman went on to tell how he scooped up several chunks of the metal from the beach and boarded the patrol boat. He tried to use his radio to summon aid, but for some unusual reason the interference was so bad he couldn't even call the three miles to his headquarters in Tacoma. When they docked at Tacoma, Jackson got first aid for his son and then reported to his superior officer, Richards, who, Jackson added to his story, didn't believe the tale. He didn't believe it until he went out to the island himself and saw the metal.

Jackson's trouble wasn't over. The next morning a mysterious visitor told Jackson to forget what he'd seen.

Later that same day the photos were developed. They showed the six objects, but the film was badly spotted and fogged, as if the film had been exposed to some kind of radiation.

Then Simpson told about his brush with mysterious callers. He said that Jackson was not alone as far as mysterious callers were concerned, the Tacoma newspapers had been getting calls from an anonymous tipster telling exactly what was going on in Simpson's hotel room. This was a very curious situation because no one except Simpson, the airline pilot, and the two harbor patrolmen knew what was taking place. The room had even been thoroughly searched for hidden microphones.

That is the way the story stood a few hours after Lieutenant Brown and Captain Davidson arrived in Tacoma.

After asking Jackson and Richards a few questions, the two intelligence agents left, reluctant even to take any of the fragments. As some writers who have since written about this incident have said, Brown and Davidson seemed to be anxious to leave and afraid to touch the fragments of the UFO, as if they knew something more about them. The two officers went to McChord AFB, near Tacoma, where their B-25 was parked, held a conference with the intelligence officer at McChord, and took off for their home base, Hamilton. When they left McChord they had a good idea as to the identity of the UFO's. Fortunately they told the McChord intelligence officer what they had determined from their interview.

In a few hours the two officers were dead. The B-25 crashed near Kelso, Washington. The crew chief and a passenger had parachuted to safety. The newspapers hinted that the airplane was sabotaged and that it was carrying highly classified material. Authorities at McChord AFB confirmed this latter point, the airplane was carrying classified material.

In a few days the newspaper publicity on the crash died down, and the Maury Island Mystery was never publicly solved.

Later reports say that the two harbor patrolmen mysteriously disappeared soon after the fatal crash.

They should have disappeared, into Puget Sound. The whole Maury Island Mystery was a hoax. The first, possibly the second-best, and the dirtiest hoax in the UFO history. One passage in the detailed official report of the Maury Island Mystery says:

Both ——— (the two harbor patrolmen) admitted that the rock fragments had nothing to do with flying saucers. The whole thing was a hoax. They had sent in the rock fragments [to a magazine publisher] as a joke. ——— One of the patrolmen wrote to ——— [the publisher] stating that the rock could have been part of a flying saucer. He had said the rock came from a flying saucer because that's what ——— [the publisher] wanted him to say.

The publisher, mentioned above, who, one of the two hoaxers said, wanted him to say that the rock fragments had come from a flying saucer, is the same one who paid the man I called Simpson $200 to investigate the case.

The report goes on to explain more details of the incident. Neither one of the two men could ever produce the photos. They "misplaced" them, they said. One of them, I forget which, was the mysterious informer who called the newspapers to report the conversations that were going on in the hotel room. Jackson's mysterious visitor didn't exist. Neither of the men was a harbor patrolman, they merely owned a couple of beat-up old boats that they used to salvage floating lumber from Puget Sound. The airplane crash was one of those unfortunate things. An engine caught on fire, burned off, and just before the two pilots could get out, the wing and tail tore off, making it impossible for them to escape. The two dead officers from Hamilton AFB smelled a hoax, accounting for their short interview and hesitancy in bothering to take the "fragments." They confirmed their convictions when they talked to the intelligence officer at McChord. It had already been established, through an informer, that the fragments were what Brown and Davidson thought, slag. The classified material on the B-25 was a file of reports the two officers offered to take back to Hamilton and had nothing to do with the Maury Island Mystery, or better, the Maury Island Hoax.

Simpson and his airline pilot friend weren't told about the hoax for one reason. As soon as it was discovered that they had been "taken," thoroughly, and were not a party to the hoax, no one wanted to embarrass them.

The majority of the writers of saucer lore have played this sighting to the hilt, pointing out as their main premise the fact that the story must be true because the government never openly exposed or prosecuted either of the two hoaxers. This is a logical premise, but a false one. The reason for the thorough investigation of the Maury Island Hoax was that the government had thought seriously of prosecuting the men. At the last minute it was decided, after talking to the two men, that the hoax was a harmless joke that had mushroomed, and that the loss of two lives and a B-25 could not be directly blamed on the two men. The story wasn't even printed because at the time of the incident, even though in this case the press knew about it, the facts were classed as evidence. By the time the facts were released they were yesterday's news. And nothing is deader than yesterday's news.

As 1947 drew to a close, the Air Force's Project Sign had outgrown its initial panic and had settled down to a routine operation. Every intelligence report dealing with the Germans' World War II aeronautical research had been studied to find out if the Russians could have developed any of the late German designs into flying saucers. Aerodynamicists at ATIC and at Wright Field's Aircraft Laboratory computed the maximum performance that could be expected from the German designs. The designers of the aircraft themselves were contacted. "Could the Russians develop a flying saucer from their designs?" The answer was, "No, there was no conceivable way any aircraft could perform that would match the reported maneuvers of the UFO's." The Air Force's Aeromedical Laboratory concurred. If the aircraft could be built, the human body couldn't stand the violent maneuvers that were reported. The aircraft-structures people seconded this, no material known could stand the loads of the reported maneuvers and heat of the high speeds.

Still convinced that the UFO's were real objects, the people at ATIC began to change their thinking. Those who were convinced that the UFO's were of Soviet origin now began to eye outer space, not because there was any evidence that the UFO's did come from outer space but because they were convinced that UFO's existed and only some unknown race with a highly developed state of technology could build such vehicles. As far as the effect on the human body was concerned, why couldn't these people, whoever they might be, stand these horrible maneuver forces? Why judge them by earthly standards? I found a memo to this effect was in the old Project Sign files.

Project Sign ended 1947 with a new problem. How do you collect interplanetary intelligence? During World War II the organization that was ATIC's forerunner, the Air Materiel Command's secret "T-2," had developed highly effective means of wringing out every possible bit of information about the technical aspects of enemy aircraft. ATIC knew these methods, but how could this be applied to spaceships? The problem was tackled with organized confusion.

If the confusion in the minds of Air Force people was organized the confusion in the minds of the public was not. Publicized statements regarding the UFO were conflicting.

A widely printed newspaper release, quoting an unnamed Air Force official in the Pentagon, said:

The "flying saucers" are one of three things:

Solar reflections on low-hanging clouds.

Small meteors that break up, their crystals catching the rays of the sun.

Icing conditions could have formed large hailstones and they might have flattened out and glided.

A follow-up, which quoted several scientists, said in essence that the unnamed Air Force official was crazy. Nobody even heard of crystallized meteors, or huge, flat hailstones, and the solar- reflection theory was absurd.

Life, *Time*, *Newsweek*, and many other news magazines carried articles about the UFO's. Some were written with tongue in cheek, others were not. All the articles mentioned the Air Force's mass- hysterical induced hallucinations. But a Veterans' Administration psychiatrist publicly pooh-poohed this. "Too many people are seeing things," he said.

It was widely suggested that all the UFO's were meteors. Two Chicago astronomers queered this. Dr. Gerard Kuiper, director of the University of Chicago observatory, was quoted as flatly saying the UFO's couldn't be meteors. "They are probably man-made," he told the Associated Press. Dr. Oliver Lee, director of Northwestern University's observatory, agreed with Dr. Kuiper and he threw in an additional confusion factor that had been in the back of many people's minds. Maybe they were our own aircraft.

The government had been denying that UFO's belonged to the U.S. from the first, but Dr. Vannevar Bush, the world-famous scientist, and Dr. Merle Tuve, inventor of the proximity fuse, added their weight. "Impossible," they said.

All of this time unnamed Air Force officials were disclaiming serious interest in the UFO subject. Yet every time a newspaper reporter went out to interview a person who had seen a UFO, intelligence agents had already been flown in, gotten the detailed story complete with sketches of the UFO, and sped back to their base to send the report to Project Sign. Many people had supposedly been "warned" not to talk too much. The Air Force was mighty interested in hallucinations.

Thus 1947 ended with various-sized question marks in the mind of the public. If you followed flying saucers closely the question mark was big, if you just noted the UFO story titles in the papers it was smaller, but it was there and it was growing. Probably none of the people, military or civilian, who had made the public statements were at all qualified to do so but they had done it, their comments had been printed, and their comments had been read. Their comments formed the question mark.

Chapter Three - The Classics

1948 was only one hour and twenty-five minutes old when a gentleman from Abilene, Texas, made the first UFO report of the year. What he saw, "a fan-shaped glow" in the sky, was insignificant as far as UFO reports go, but it ushered in a year that was to bring feverish activity to Project Sign.

With the Soviets practically eliminated as a UFO source, the idea of interplanetary spaceships was becoming more popular. During 1948 the people in ATIC were openly discussing the possibility of interplanetary visitors without others tapping their heads and looking smug. During 1948 the novelty of UFO's had worn off for the press and every John and Jane Doe who saw one didn't make the front pages as in 1947. Editors were becoming hardened, only a few of the best reports got any space. Only "The Classics" rated headlines. "The Classics" were three historic reports that were the highlights of 1948. They are called "The Classics," a name given them by the Project Blue Book staff, because: (1) they are classic examples of how the true facts of a UFO report can be twisted and warped by some writers to prove their point, (2) they are the most highly publicized reports of this early era of the UFO's, and (3) they "proved" to ATIC's intelligence specialists that UFO's were real.

The apparent lack of interest in UFO reports by the press was not a true indication of the situation. I later found out, from talking to writers, that all during 1948 the interest in UFO's was running high. The Air Force Press Desk in the Pentagon was continually being asked what progress was being made in the UFO investigation. The answer was, "Give us time. This job can't be done in a week." The press respected this and was giving them time. But every writer worth his salt has contacts, those "usually reliable sources" you read about, and these contacts were talking. All during 1948 contacts in the Pentagon were telling how UFO reports were rolling in at the rate of several per day and how ATIC UFO investigation teams were flying out of Dayton to investigate them. They were telling how another Air Force investigative organization had been called in to lighten ATIC's load and allow ATIC to concentrate on the analysis of the reports. The writers knew this was true because they had crossed paths with these men whom they had mistakenly identified as FBI agents. The FBI was never officially interested in UFO sightings. The writers' contacts in the airline industry told about the UFO talk from V.P.'s down to the ramp boys. Dozens of good, solid, reliable, experienced airline pilots were seeing UFO's. All of this led to one conclusion: whatever the Air Force had to say, when it was ready to talk, would be newsworthy. But the Air Force wasn't ready to talk.

Project Sign personnel were just getting settled down to work after the New Year's holiday when the "ghost rockets" came back to the Scandinavian countries of Europe. Air attaches in Sweden, Denmark, and Norway fired wires to ATIC telling about the reports. Wires went back asking for more information.

The "ghost rockets," so tagged by the newspapers, had first been seen in the summer of 1946, a year before the first UFO sighting in the U.S. There were many different descriptions for the reported objects. They were usually seen in the hours of darkness and almost always traveling at extremely high speeds. They were shaped like a ball or projectile, were a bright green, white, red, or yellow and sometimes made sounds. Like their American cousins, they were always so far away that no details could be seen. For no good reason, other than speculation and circulation, the newspapers had soon begun to refer authoritatively to these "ghost rockets" as guided missiles, and implied that they were from Russia. Peenemunde, the great German missile development center and birthplace of the V-l and V-2 guided missiles, came in for its share of suspicion since it was held by the Russians. By the end of the summer of 1946 the reports were widespread, coming from Denmark, Norway, Spain, Greece, French Morocco, Portugal, and Turkey. In 1947, after no definite conclusions as to identity of the "rockets" had been established, the reports died out. Now in early January 1948 they broke out again. But Project Sign personnel were too busy to worry about European UFO reports, they were busy at home. A National Guard pilot had just been killed chasing a UFO.

On January 7 all of the late papers in the U.S. carried headlines similar to those in the Louisville *Courier*: "F-51 and Capt. Mantell Destroyed Chasing Flying Saucer." This was Volume I of "The Classics," the Mantell Incident.

At one-fifteen on that afternoon the control tower operators at Godman AFB, outside Louisville, Kentucky, received a telephone call from the Kentucky State Highway Patrol. The patrol wanted to know if Godman Tower knew anything about any unusual aircraft in the vicinity. Several people from Maysville, Kentucky, a small town 80 miles east of Louisville, had reported seeing a strange aircraft. Godman knew that they had nothing in the vicinity so they called Flight Service at Wright-Patterson AFB. In a few minutes Flight Service called back. Their air Traffic control board showed no flights in the area. About twenty minutes later the state police called again. This time people from the towns of Owensboro and Irvington, Kentucky, west of Louisville, were reporting a strange craft. The report from these two towns was a little more complete. The townspeople had described the object to the state police as being "circular, about 250 to 300 feet in diameter," and moving westward at a "pretty good clip." Godman Tower checked Flight Service again. Nothing. All this time the tower operators had been looking for the reported object. They theorized that since the UFO had had to pass north of Godman to get from Maysville to Owensboro it might come back.

At one forty-five they saw it, or something like it. Later, in his official report, the assistant tower operator said that he had seen the object for several minutes before he called his chiefs attention to it. He said that he had been reluctant to "make a flying saucer report." As soon as the two men in the tower had assured themselves that the UFO they saw was not an airplane or a weather balloon, they called Flight Operations. They wanted the operations officer to see the UFO. Before long word of the sighting had gotten around to key personnel on the base, and several officers, besides the base operations officer and the base intelligence officer, were in the tower. All of them looked at the UFO through the tower's 6 x 50 binoculars and decided they couldn't identify it. About this time Colonel Hix, the base commander, arrived. He looked and he was baffled. At two-thirty, they reported, they were discussing what should be done when four F-51's came into view, approaching the base from the south.

The tower called the flight leader, Captain Mantell, and asked him to take a look at the object and try to identify it. One F-51 in the flight was running low on fuel, so he asked permission to go on to his base. Mantell took his two remaining wing men, made a turn, and started after the UFO. The people in Godman Tower were directing him as none of the pilots could see the object at this time. They gave Mantell an initial heading toward the south and the flight was last seen heading in the general direction of the UFO.

By the time the F-51's had climbed to 10,000 feet, the two wing men later reported, Mantell had pulled out ahead of them and they could just barely see him. At two forty-five Mantell called the tower and said, "I see something above and ahead of me and I'm still climbing." All the people in the tower heard Mantell say this and they heard one of the wing men call back and ask, "What the hell are we looking for?" The tower immediately called Mantell and asked him for a description of what he saw. Odd as it may seem, no one can remember exactly what he answered. Saucer historians have credited him with saying, "I've sighted the thing. It looks metallic and it's tremendous in size. . . . Now it's starting to climb." Then in a few seconds he is supposed to have called and said, "It's above me and I'm gaining on it. I'm going to 20,000 feet." Everyone in the tower agreed on this one last bit of the transmission, "I'm going to 20,000 feet," but didn't agree on the first part, about the UFO's being metallic and tremendous.

The two wing men were now at 15,000 feet and trying frantically to call Mantell. He had climbed far above them by this time and was out of sight. Since none of them had any oxygen they were worried about Mantell. Their calls were not answered. Mantell never talked to anyone again. The two wing men leveled off at 15,000 feet, made another fruitless effort to call Mantell, and started to come back down. As they passed Godman Tower on their way to their base, one of them said something to the effect that all he had seen was a reflection on his canopy.

When they landed at their base, Standiford Field, just north of Godman, one pilot had his F-51 refueled and serviced with oxygen, and took off to search the area again. He didn't see anything.

At three-fifty the tower lost sight of the UFO. A few minutes later they got word that Mantell had crashed and was dead.

Several hours later, at 7:20P.M., airfield towers all over the Midwest sent in frantic reports of another UFO. In all about a dozen airfield towers reported the UFO as being low on the southwestern horizon and disappearing after about twenty minutes. The writers of saucer lore say this UFO was what Mantell was chasing when he died; the Air Force says *this* UFO was Venus.

The people on Project Sign worked fast on the Mantell Incident. Contemplating a flood of queries from the press as soon as they heard about the crash, they realized that they had to get a quick answer. Venus had been the target of a chase by an Air Force F-51 several weeks before and there were similarities between this sighting and the Mantell Incident. So almost before the rescue crews had reached the crash, the word "Venus" went out. This satisfied the editors, and so it stood for about a year; Mantell had unfortunately been killed trying to reach the planet Venus.

To the press, the nonchalant, offhand manner with which the sighting was written off by the Air Force public relations officer showed great confidence in the conclusion, Venus, but behind the barbed-wire fence that encircled ATIC the nonchalant attitude didn't exist among the intelligence analysts. One man had already left for Louisville and the rest were doing some tall speculating. The story about the tower-to-air talk. "It looks metallic and it's tremendous in size," spread fast. Rumor had it that the tower had carried on a running conversation with the pilots and that there was more information than was so far known. Rumor also had it that this conversation had been recorded. Unfortunately neither of these rumors was true.

Over a period of several weeks the file on the Mantell Incident grew in size until it was the most thoroughly investigated sighting of that time, at least the file was the thickest.

About a year later the Air Force released its official report on the incident. To use a trite term, it was a masterpiece in the art of "weasel wording." It said that the UFO might have been Venus or it could have been a balloon. Maybe two balloons. It probably was Venus except that this is doubtful because Venus was too dim to be seen in the afternoon. This jolted writers who had been following the UFO story. Only a few weeks before, *The Saturday Evening Post* had published a two-part story entitled "What You Can Believe about Flying Saucers." The story had official sanction and had quoted the Venus theory as a positive solution. To clear up the situation, several writers were allowed to interview a major in the Pentagon, who was the Air Force's Pentagon "expert" on UFO's. The major was asked directly about the conclusion of the Mantell Incident, and he flatly stated that it was Venus. The writers pointed out the official Air Force analysis. The major's answer was, "They checked again and it was Venus." He didn't know who "they" were, where they had checked, or what they had checked, but it was Venus. The writers then asked, "If there was a later report they had made why wasn't it used as a conclusion?" "Was it available?" The answer to the last question was "No," and the lid snapped back down. This interview gave the definite impression that the Air Force was unsuccessfully trying to cover up some very important information, using Venus as a front. Nothing excites a newspaper or magazine writer more than to think he has stumbled onto a big story and that someone is trying to cover it up. Many writers thought this after the interview with the major, and many still think it. You can't really blame them, either.

In early 1952 I got a telephone call on ATIC's direct line to the Pentagon. It was a colonel in the Director of Intelligence's office. The Office of Public Information had been getting a number of queries about all of the confusion over the Mantell Incident. What was the answer?

I dug out the file. In 1949 all of the original material on the incident had been microfilmed, but something had been spilled on the film. Many sections were so badly faded they were illegible. As I had to do with many of the older sightings that were now history, I collected what I could from the file, filling in the blanks by talking to people who had been at ATIC during the early UFO era. Many of these people were still around, "Red" Honnacker, George Towles, Al Deyarmond, Nick Post, and many others. Most of them were civilians, the military had been transferred out by this time.

Some of the press clippings in the file mentioned the Pentagon major and his concrete proof of Venus. I couldn't find this concrete proof in the file so I asked around about the major. The major, I found, was an officer in the Pentagon who had at one time written a short intelligence summary about UFO's. He had never been stationed at ATIC, nor was he especially well versed on the UFO problem. When the word of the press conference regarding the Mantell Incident came down, a UFO expert was needed. The major, because of his short intelligence summary on UFO's, became the "expert." He had evidently conjured up "they" and "their later report" to support his Venus answer because the writers at the press conference had him in a corner. I looked farther.

Fortunately the man who had done the most extensive work on the incident, Dr. J. Allen Hynek, head of the Ohio State University Astronomy Department, could be contacted. I called Dr. Hynek and arranged to meet him the next day.

Dr. Hynek was one of the most impressive scientists I met while working on the UFO project, and I met a good many. He didn't do two things that some of them did: give you the answer before he knew the question; or immediately begin to expound on his accomplishments in the field of science. I arrived at Ohio State just before lunch, and Dr. Hynek invited me to eat with him at the faculty club. He wanted to refer to some notes he had on the Mantell Incident and they were in his office, so we discussed UFO's in general during lunch.

Back in his office he started to review the Mantell Incident. He had been responsible for the weasel-worded report that the Air Force released in late 1949, and he apologized for it. Had he known that it was going to cause so much confusion, he said, he would have been more specific. He thought the incident was a dead issue. The reason that Venus had been such a strong suspect was that it was in almost the same spot in the sky as the UFO. Dr. Hynek referred to his notes and told me that at 3:00P.M., Venus had been south southwest of Godman and 33 degrees above the southern horizon. At 3:00P.M. the people in the tower estimated the UFO to be southwest of Godman and at an elevation of about 45 degrees. Allowing for human error in estimating directions and angles, this was close. I agreed. There was one big flaw in the theory, however. Venus wasn't bright enough to be seen. He had computed the brilliance of the planet, and on the day in question it was only six times as bright as the surrounding sky. Then he explained what this meant. Six times may sound like a lot, but it isn't. When you start looking for a pinpoint of light only six times as bright as the surrounding sky, it's almost impossible to find it, even on a clear day.

Dr. Hynek said that he didn't think that the UFO was Venus.

I later found out that although it was a relatively clear day there was considerable haze.

I asked him about some of the other possibilities. He repeated the balloon, canopy-reflection, and sundog theories but he refused to comment on them since, as he said, he was an astrophysicist and would care to comment only on the astrophysical aspects of the sightings.

I drove back to Dayton convinced that the UFO wasn't Venus. Dr. Hynek had said Venus would have been a pinpoint of light. The people in the tower had been positive of their descriptions, their statements brought that out. They couldn't agree on a description, they called the UFO "a parachute," "an ice cream cone tipped with red," "round and white," "huge and silver or metallic," "a small white object," "one fourth the size of the full moon," but all the descriptions plainly indicated a large object. None of the descriptions could even vaguely be called a pinpoint of light.

This aspect of a definite shape seemed to eliminate the sundog theory too. Sundogs, or parhelia, as they are technically known, are caused by ice particles reflecting a diffused light. This would not give a sharp outline. I also recalled two instances where Air Force pilots had chased sundogs. In both instances when the aircraft began to climb, the sundog disappeared. This was because the angle of reflection changed as the airplane climbed several thousand feet. These sundog-caused UFO's also had fuzzy edges.

I had always heard a lot of wild speculation about the condition of Mantell's crashed F-51, so I wired for a copy of the accident report. It arrived several days after my visit with Dr. Hynek. The report said that the F-51 had lost a wing due to excessive speed in a dive after Mantell had "blacked out" due to the lack of oxygen. Mantell's body had not burned, not disintegrated, and was not full of holes; the wreck was not radioactive, nor was it magnetized.

One very important and pertinent question remained. Why did Mantell, an experienced pilot, try to go to 20,000 feet when he didn't even have an oxygen mask? If he had run out of oxygen, it would have been different Every pilot and crewman has it pounded into him, "Do not, under any circumstances, go above 15,000 feet without oxygen." In high-altitude indoctrination during World War II, I made several trips up to 30,000 feet in a pressure chamber. To demonstrate anoxia we would leave our oxygen masks off until we became dizzy. A few of the more hardy souls could get to 15,000 feet, but nobody ever got over 17,000. Possibly Mantell thought he could climb up to 20,000 in a hurry and get back down before he got anoxia and blacked out, but this would be a foolish chance. This point was covered in the sighting report. A long-time friend of Mantell's went on record as saying that he'd flown with him several years and knew him personally. He couldn't conceive of Mantell's even thinking about disregarding his lack of oxygen. Mantell was one of the most cautious pilots he knew. "The only thing I can think," he commented, "was that he was after something that he believed to be more important than his life or his family."

My next step was to try to find out what Mantell's wing men had seen or thought but this was a blind alley. All of this evidence was in the ruined portion of the microfilm, even their names were missing. The only reference I could find to them was a vague passage indicating they hadn't seen anything.

I concentrated on the canopy-reflection theory. It is widely believed that many flying saucers appear to pilots who are actually chasing a reflection on their canopy. I checked over all the reports we had on file. I couldn't find one that had been written off for this reason. I dug back into my own flying experience and talked to a dozen pilots. All of us had momentarily been startled by a reflection on the aircraft's canopy or wing, but in a second or two it had been obvious that it was a reflection. Mantell chased the object for at least fifteen to twenty minutes, and it is inconceivable that he wouldn't realize in that length of time that he was chasing a reflection.

About the only theory left to check was that the object might have been one of the big, 100-foot-diameter, "skyhook" balloons. I rechecked the descriptions of the UFO made by the people in the tower. The first man to sight the object called it a parachute; others said ice cream cone, round, etc. All of these descriptions fit a balloon. Buried deep in the file were two more references to balloons that I had previously missed. Not long after the object had disappeared from view at Godman AFB, a man from Madisonville, Kentucky, called Flight Service in Dayton. He had seen an object traveling southeast. He had looked at it through a telescope and it was a balloon. At four forty-five an astronomer living north of Nashville, Tennessee, called in. He had also seen a UFO, looked at it through a telescope, and it was a balloon.

In the thousands of words of testimony and evidence taken on the Mantell Incident this was the only reference to balloons. I had purposely not paid too much attention to this possibility because I was sure that it had been thoroughly checked back in 1948. Now I wasn't sure.

I talked with one of the people who had been in on the Mantell investigation. The possibility of a balloon's causing the sighting had been mentioned but hadn't been followed up for two reasons. Number one was that everybody at ATIC was convinced that the object Mantell was after was a spaceship and that this was the only course they had pursued. When the sighting grew older and no spaceship proof could be found, everybody jumped on the Venus band wagon, as this theory had "already been established." It was an easy way out. The second reason was that a quick check had been made on weather balloons and none were in the area. The big skyhook balloon project was highly classified at that time, and since they were all convinced that the object was of interplanetary origin (a minority wanted to give the Russians credit), they didn't want to bother to buck the red tape of security to get data on skyhook flights.

The group who supervise the contracts for all the skyhook research flights for the Air Force are located at Wright Field, so I called them. They had no records on flights in 1948 but they did think that the big balloons were being launched from Clinton County AFB in southern Ohio at that time. They offered to get the records of the winds on January 7 and see what flight path a balloon launched in southwestern Ohio would have taken. In a few days they had the data for me.

Unfortunately the times of the first sightings, from the towns outside Louisville, were not exact but it was possible to partially reconstruct the sequence of events. The winds were such that a skyhook balloon launched from Clinton County AFB could be seen from the town east of Godman AFB, the town from which the first UFO was reported to the Kentucky State Police. It is not unusual to be able to see a large balloon for 50 to 60 miles. The balloon could have traveled west for a while, climbing as it moved with the strong east winds that were blowing that day and picking up speed as the winds got stronger at altitude. In twenty minutes it could have been in a position where it could be seen from Owensboro and Irvington, Kentucky, the two towns west of Godman. The second reports to the state police had come from these two towns. Still climbing, the balloon would have reached a level where a strong wind was blowing in a southerly direction. The jet-stream winds were not being plotted in 1948 but the weather chart shows strong indications of a southerly bend in the jet stream for this day. Jet stream or not, the balloon would have moved rapidly south, still climbing. At a point somewhere south or southwest of Godman it would have climbed through the southerly-moving winds to a calm belt at about 60,000 feet. At this level it would slowly drift south or southeast. A skyhook balloon can be seen at 60,000.

When first seen by the people in Godman Tower, the UFO was south of the air base. It was relatively close and looked "like a parachute," which a balloon does. During the two hours that it was in sight, the observers reported that it seemed to hover, yet each observer estimated the time he looked at the object through the binoculars and timewise the descriptions ran "huge," "small," "one fourth the size of a full moon," "one tenth the size of a full moon." Whatever the UFO was, it was slowly moving away. As the balloon continued to drift in a southerly direction it would have picked up stronger winds, and could have easily been seen by the astronomers in Madisonville, Kentucky, and north of Nashville an hour after it disappeared from view at Godman.

Somewhere in the archives of the Air Force or the Navy there are records that will show whether or not a balloon was launched from Clinton County AFB, Ohio, on January 7, 1948. I never could find these records. People who were working with the early skyhook projects "remember" operating out of Clinton County AFB in 1947 but refuse to be pinned down to a January 7 flight. Maybe, they said.

The Mantell Incident is the same old UFO jigsaw puzzle. By assuming the shape of one piece, a balloon launched from southwestern Ohio, the whole picture neatly falls together. It shows a huge balloon that Captain Thomas Mantell died trying to reach. He didn't know that he was chasing a balloon because he had never heard of a huge, 100-foot- diameter skyhook balloon, let alone seen one. Leave out the one piece of the jigsaw puzzle and the picture is a UFO, "metallic and tremendous in size."

It *could* have been a balloon. This is the answer I phoned back to the Pentagon.

During January and February of 1948 the reports of "ghost rockets" continued to come from air attaches in foreign countries near the Baltic Sea. People in North Jutland, Norway, Denmark, Sweden, and Germany reported "balls of fire traveling slowly across the sky." The reports were very sketchy and incomplete, most of them accounts from newspapers. In a few days the UFO's were being seen all over Europe and South America. Foreign reports hit a peak in the latter part of February and U.S. newspapers began to pick up the stories.

The Swedish Defense Staff supposedly conducted a comprehensive study of the incidents and concluded that they were all explainable in terms of astronomical phenomena. Since this was UFO history, I made several attempts to get some detailed and official information on this report and the sightings, but I was never successful.

The ghost rockets left in March, as mysteriously as they had arrived.

All during the spring of 1948 good reports continued to come in. Some were just run-of-the-mill but a large percentage of them were good, coming from people whose reliability couldn't be questioned. For example, three scientists reported that for thirty seconds they had watched a round object streak across the sky in a highly erratic flight path near the Army's secret

White Sands Proving Ground. And on May 28 the crew of an Air Force C-47 had three UFO's barrel in from "twelve o'clock high" to buzz their transport.

On July 21 a curious report was received from the Netherlands. The day before several persons reported seeing a UFO through high broken clouds over The Hague. The object was rocket-shaped, with two rows of windows along the side. It was a poor report, very sketchy and incomplete, and it probably would have been forgotten except that four nights later a similar UFO almost collided with an Eastern Airlines DC-3. This near collision is Volume II of "The Classics."

On the evening of July 24, 1948, an Eastern Airlines DC-3 took off from Houston, Texas. It was on a scheduled trip to Atlanta, with intermediate stops in between. The pilots were Clarence S. Chiles and John B. Whitted. At about 2:45 A.M., when the flight was 20 miles southwest of Montgomery, the captain, Chiles, saw a light dead ahead and closing fast. His first reaction, he later reported to an ATIC investigation team, was that it was a jet, but in an instant he realized that even a jet couldn't close as fast as this light was closing. Chiles said he reached over, gave Whitted, the other pilot, a quick tap on the arm, and pointed. The UFO was now almost on top of them. Chiles racked the DC-3 into a tight left turn. Just as the UFO flashed by about 700 feet to the right, the DC-3 hit turbulent air. Whitted looked back just as the UFO pulled up in a steep climb.

Both the pilots had gotten a good look at the UFO and were able to give a good description to the Air Force intelligence people. It was a B-29 fuselage. The underside had a "deep blue glow." There were "two rows of windows from which bright lights glowed," and a "50-foot trail of orange-red flame" shot out the back.

Only one passenger was looking out of the window at the time. The ATIC investigators talked to him. He said he saw a "strange, eerie streak of light, very intense," but that was all, no details. He said that it all happened before he could adjust his eyes to the darkness.

Minutes later a crew chief at Robins Air Force Base in Macon, Georgia, reported seeing an extremely bright light pass overhead, traveling at a high speed. A few days later another report from the night of July 24 came in. A pilot, flying near the Virginia-North Carolina state line, reported that he had seen a "bright shooting star" in the direction of Montgomery, Alabama, at about the exact time the Eastern Airlines DC-3 was "buzzed."

According to the old timers at ATIC, this report shook them worse than the Mantell Incident. This was the first time two reliable sources had been really close enough to anything resembling a UFO to get a good look and live to tell about it. A quick check on a map showed that the UFO that nearly collided with the airliner would have passed almost over Macon, Georgia, after passing the DC-3. It had been turning toward Macon when last seen. The story of the crew chief at Robins AFB, 200 miles away, seemed to confirm the sighting, not to mention the report from near the Virginia-North Carolina state line.

In intelligence, if you have something to say about some vital problem you write a report that is known as an "Estimate of the Situation." A few days after the DC-3 was buzzed, the people at ATIC decided that the time had arrived to make an Estimate of the Situation. The situation was the UFO's; the estimate was that they were interplanetary!

It was a rather thick document with a black cover and it was printed on legal-sized paper. Stamped across the front were the words TOP SECRET.

It contained the Air Force's analysis of many of the incidents I have told you about plus many similar ones. All of them had come from scientists, pilots, and other equally credible observers, and each one was an unknown.

The document pointed out that the reports hadn't actually started with the Arnold Incident. Belated reports from a weather observer in Richmond, Virginia, who observed a "silver disk" through his theodolite telescope; an F-47 pilot and three pilots in his formation who saw a "silver flying wing," and the English "ghost airplanes" that had been picked up on radar early in 1947 proved this point. Although reports on them were not received until after the Arnold sighting, these incidents all had taken place earlier.

When the estimate was completed, typed, and approved, it started up through channels to higher-command echelons. It drew considerable comment but no one stopped it on its way up.

A matter of days after the Estimate of the Situation was signed, sealed, and sent on its way, the third big sighting of 1948, Volume III of "The Classics," took place. The date was October 1, and the place was Fargo, North Dakota; it was the famous Gorman Incident, in which a pilot fought a "duel of death" with a UFO.

The pilot was George F. Gorman, a twenty-five-year-old second lieutenant in the North Dakota Air National Guard.

It was eight-thirty in the evening and Gorman was coming into Fargo from a cross-country flight. He flew around Fargo for a while and about nine o'clock decided to land. He called the control tower for landing instructions and was told that a Piper Cub was in the area. He saw the Cub below him. All of a sudden what appeared to be the taillight of another airplane passed him on his right. He called the tower and complained but they assured him that no other aircraft except the Cub were in the area. Gorman could still see the light so he decided to find out what it was. He pushed the F-51 over into a turn and cut in toward the light. He could plainly see the Cub outlined against the city lights below, but he could see no outline of a body near the mysterious light. He gave the '51 more power and closed to within a 1,000 yards, close enough to estimate that the light was 6 to 8 inches in diameter, was sharply outlined, and was blinking on and off. Suddenly the light became steady as it apparently put on power; it pulled into a sharp left bank and made a pass at the tower. The light zoomed up with the F-51 in hot pursuit. At 7,000 feet it made a turn. Gorman followed and tried to cut inside the light's turn to get closer to it but he couldn't do it. The light made another turn, and this time the '51 closed on a collision course. The UFO appeared to try to ram the '51, and Gorman had to dive to get out of the way. The UFO passed over the '51's canopy with only a few feet to spare. Again both the F-51 and the object turned and closed on each other head on, and again the pilot had to dive out to prevent a collision. All of a sudden the light began to climb and disappeared.

"I had the distinct impression that its maneuvers were controlled by thought or reason," Gorman later told ATIC investigators.

Four other observers at Fargo partially corroborated his story, an oculist, Dr. A. D. Cannon, the Cub's pilot, and his passenger, Einar Neilson. They saw a light "moving fast," but did not witness all the maneuvers that Gorman reported. Two CAA employees on the ground saw a light move over the field once.

Project Sign investigators rushed to Fargo. They had wired ahead to ground the plane. They wanted to check it over before it flew again. When they arrived, only a matter of hours after the incident, they went over the airplane, from the prop spinner to the rudder trim tab, with a Geiger counter. A chart in the official report shows where every Geiger counter reading was taken. For comparison they took readings on a similar airplane that hadn't been flown for several days. Gorman's airplane was more radioactive. They rushed around, got sworn statements from the tower operators and oculist, and flew back to Dayton.

In the file on the Gorman Incident I found an old memo reporting the meeting that was held upon the ATIC team's return from Fargo. The memo concluded that some weird things were taking place.

The historians of the UFO agree. Donald Keyhoe, a retired Marine Corps major and a professional writer, author of *The Flying Saucers Are Real* and *Flying Saucers from Outer Space*, needles the Air Force about the Gorman Incident, pointing out how, after feebly hinting that the light could have been a lighted weather balloon, they dropped it like a hot UFO. Some person by the name of Wilkins, in an equally authoritative book, says that the Gorman Incident "stumped" the Air Force. Other assorted historians point out that normally the UFO's are peaceful, Gorman and Mantell just got too inquisitive, "they" just weren't ready to be observed closely. If the Air Force hadn't slapped down the security lid, these writers might not have reached this conclusion. There have been other and more lurid "duels of death."

On June 21, 1952, at 10:58P.M., a Ground Observer Corps spotter reported that a slow-moving craft was nearing the AEC's Oak Ridge Laboratory, an area so secret that it is prohibited to aircraft. The spotter called the light into his filter center and the filter center relayed the message to the ground control intercept radar. They had a target. But before they could do more than confirm the GOC spotter's report, the target faded from the radarscope.

An F-47 aircraft on combat air patrol in the area was vectored in visually, spotted a light, and closed on it. They "fought" from 10,000 to 27,000 feet, and several times the object made what seemed to be ramming attacks. The light was described as white, 6 to 8 inches in diameter, and blinking until it put on power. The pilot could see no silhouette around the light. The similarity to the Fargo case was striking.

On the night of December 10, 1952, near another atomic installation, the Hanford plant in Washington, the pilot and radar observer of a patrolling F-94 spotted a light while flying at 26,000 feet. The crew called their ground control station and were told that no planes were known to be in the area. They closed on the object and saw a large, round, white "thing" with a dim reddish light coming from two "windows." They lost visual contact, but got a radar lock-on. They reported that when they attempted to close on it again it would reverse direction and dive away. Several times the plane altered course itself because collision seemed imminent.

In each of these instances, as well as in the case narrated next, the sources of the stories were trained airmen with excellent reputations. They were sincerely baffled by what they had seen. They had no conceivable motive for falsifying or "dressing up" their reports.

The other dogfight occurred September 24, 1952, between a Navy pilot of a TBM and a light over Cuba.

The pilot had just finished making some practice passes for night fighters when he spotted an orange light to the east of his plane. He checked on aircraft in the area, learned that the object was unidentified, and started after it. Here is his report, written immediately after he landed:

As it [the light] approached the city from the east it started a left turn. I started to intercept. During the first part of the chase the closest I got to the light was 8 to 10 miles. At this time it appeared to be as large as an SNJ and had a greenish tail that looked to be five to six times as long as the light's diameter. This tail was seen several times in the next 10 minutes in periods of from 5 to 30 seconds each. As I reached 10,000 feet it appeared to be at 15,000 feet and in a left turn. It took 40 degrees of bank to keep the nose of my plane on the light. At this time I estimated the light to be in a 10-to-15-mile orbit.

At 12,000 feet I stopped climbing, but the light was still climbing faster than I was. I then reversed my turn from left to right and the light also reversed. As I was not gaining distance, I held a steady course south trying to estimate a perpendicular between the light and myself. The light was moving north, so I turned north. As I turned, the light appeared to move west, then south over the base. I again tried to intercept but the light appeared to climb rapidly at a 60- degree angle. It climbed to 35,000 feet, then started a rapid descent.

Prior to this, while the light was still at approximately 15,000 feet, I deliberately placed it between the moon and myself three times to try to identify a solid body. I and my two crewmen all had a good view of the light as it passed the moon. We could see no solid body. We considered the fact that it might be an aerologist's balloon, but we did not see a silhouette. Also, we would have rapidly caught up with and passed a balloon.

During its descent, the light appeared to slow down at about 10,000 feet, at which time I made three runs on it. Two were on a 90-degree collision course, and the light traveled at tremendous speed across my bow. On the third run I was so close that the light blanked out the airfield below me. Suddenly it started a dive and I followed, losing it at 1,500 feet.

In *this* incident the UFO *was* a balloon.

The following night a lighted balloon was sent up and the pilot was ordered up to compare his experiences. He duplicated his dogfight— illusions and all. The Navy furnished us with a long analysis of the affair, explaining how the pilot had been fooled.

In the case involving the ground observer and the F-47 near the atomic installation, we plotted the winds and calculated that a lighted balloon was right at the spot where the pilot encountered the light.

In the other instance, the "white object with two windows," we found that a skyhook balloon had been plotted at the exact site of the "battle."

Gorman fought a lighted balloon too. An analysis of the sighting by the Air Weather Service sent to ATIC in a letter dated January 24, 1949, proved it. The radioactive F-51 was decontaminated by a memo from a Wright Field laboratory explaining that a recently flown airplane will be more radioactive than one that has been on the ground for several days. An airplane at 20,000 to 30,000 feet picks up more cosmic rays than one shielded by the earth's ever present haze.

Why can't experienced pilots recognize a balloon when they see one? If they are flying at night, odd things can happen to their vision. There is the problem of vertigo as well as disorientation brought on by flying without points of reference. Night fighters have told dozens of stories of being fooled by lights.

One night during World War II we had just dumped a load of bombs on a target when a "night fighter" started to make a pass at us. Everyone in the cockpit saw the fighter's red-hot exhaust stack as he bore down on us. I cut loose with six caliber-.50 machine guns. Fortunately I missed the "night fighter"—if I'd have shot it I'd have fouled up the astronomers but good because the "night fighter" was Venus.

While the people on Project Sign were pondering over Lieutenant Gorman's dogfight with the UFO—at the time they weren't even considering the balloon angle—the Top Secret Estimate of the Situation was working its way up into the higher echelons of the Air Force. It got to the late General Hoyt S. Vandenberg, then Chief of Staff, before it was batted back down. The general wouldn't buy interplanetary vehicles. The report lacked proof. A group from ATIC went to the Pentagon to bolster their position but had no luck, the Chief of Staff just couldn't be convinced.

The estimate died a quick death. Some months later it was completely declassified and relegated to the incinerator. A few copies, one of which I saw, were kept as mementos of the golden days of the UFO's.

The top Air Force command's refusal to buy the interplanetary theory didn't have any immediate effect upon the morale of Project Sign because the reports were getting better.

A belated report that is more of a collectors' item than a good UFO sighting came into ATIC in the fall of 1948. It was from Moscow. Someone, I could never find out exactly who, reported a huge "smudge-like" object in the sky.

Then radar came into the picture. For months the anti-saucer factions had been pointing their fingers at the lack of radar reports, saying, "If they exist, why don't they show up on radarscopes?" When they showed up on radarscopes, the UFO won some converts.

On October 15 an F-61, a World War II "Black Widow" night fighter, was on patrol over Japan when it picked up an unidentified target on its radar. The target was flying between 5,000 and 6,000 feet and traveling about 200 miles per hour. When the F-61 tried to intercept it would get to within 12,000 feet of the UFO only to have it accelerate to an estimated 1,200 miles per hour, leaving the F-61 far behind before slowing down again. The F-61 crew made six attempts to close on the UFO. On one pass, the crew said, they did get close enough to see its silhouette. It was 20 to 30 feet long and looked "like a rifle bullet."

Toward the end of November a wire came into Project Sign from Germany. It was the first report where a UFO was seen and simultaneously picked up on radar. This type of report, the first of many to come, is one of the better types of UFO reports. The wire said:

At 2200 hours, local time, 23 November 1948, Capt. ——— saw an object in the air directly east of this base. It was at an unknown altitude. It looked like a reddish star and was moving in a southerly direction across Munich, turning slightly to the southwest then the southeast. The speed could have been between 200 to 600 mph, the actual speed could not be estimated, not knowing the height. Capt. —- —- called base operations and they called the radar station. Radar reported that they had seen nothing on their scope but would check again. Radar then called operations to report that they did have a target at 27,000 feet, some 30 miles south of Munich, traveling at 900 mph. Capt. ——— reported that the object that he saw was now in that area. A few minutes later radar called again to say that the target had climbed to 50,000 feet, and was circling 40 miles south of Munich.

Capt. ——— is an experienced pilot now flying F-80's and is considered to be completely reliable. The sighting was verified by Capt. ——— , also an F-80 pilot.

The possibility that this was a balloon was checked but the answer from Air Weather Service was "not a balloon." No aircraft were in the area. Nothing we know of, except possibly experimental aircraft, which are not in Germany, can climb 23,000 feet in a matter of minutes and travel 900 miles per hour.

By the end of 1948, Project Sign had received several hundred UFO reports. Of these, 167 had been saved as good reports. About three dozen were

"Unknown." Even though the UFO reports were getting better and more numerous, the enthusiasm over the interplanetary idea was cooling off. The same people who had fought to go to Godman AFB to talk to Colonel Hix and his UFO observers in January now had to be prodded when a sighting needed investigating. More and more work was being pushed off onto the other investigative organization that was helping ATIC. The kickback on the Top Secret Estimate of the Situation was beginning to dampen a lot of enthusiasms. It was definitely a bear market for UFO's.

A bull market was on the way, however. Early 1949 was to bring "little lights" and green fireballs.

The "little lights" were UFO's, but the green fireballs were real.

Chapter Four - Green Fireballs, Project Twinkle, Little Lights, and Grudge

At exactly midnight on September 18, 1954, my telephone rang. It was Jim Phalen, a friend of mine from the Long Beach *Press-Telegram*, and he had a "good flying saucer report," hot off the wires. He read it to me. The lead line was: "Thousands of people saw a huge fireball light up dark New Mexico skies tonight."

The story went on to tell about how a "blinding green" fireball the size of a full moon had silently streaked southeast across Colorado and northern New Mexico at eight-forty that night. Thousands of people had seen the fireball. It had passed right over a crowded football stadium at Santa Fe, New Mexico, and people in Denver said it "turned night into day." The crew of a TWA airliner flying into Albuquerque from Amarillo, Texas, saw it. Every police and newspaper switchboard in the two-state area was jammed with calls.

One of the calls was from a man inquiring if anything unusual had happened recently. When he was informed about the mysterious fireball he heaved an audible sigh of relief, "Thanks," he said, "I was afraid I'd gotten some bad bourbon." And he hung up.

Dr. Lincoln La Paz, world-famous authority on meteorites and head of the University of New Mexico's Institute of Meteoritics, apparently took the occurrence calmly. The wire story said he had told a reporter that he would plot its course, try to determine where it landed, and go out and try to find it. "But," he said, "I don't expect to find anything."

When Jim Phalen had read the rest of the report he asked, "What was it?"

"It sounds to me like the green fireballs are back," I answered.

"What the devil are green fireballs?"

What the devil *are* green fireballs? I'd like to know. So would a lot of other people.

The green fireballs streaked into UFO history late in November 1948, when people around Albuquerque, New Mexico, began to report seeing mysterious "green flares" at night. The first reports mentioned only a "green streak in the sky," low on the horizon. From the description the Air Force Intelligence people at Kirtland AFB in Albuquerque and the Project Sign people at ATIC wrote the objects off as flares. After all, thousands of GI's had probably been discharged with a duffel bag full of "liberated" Very pistols and flares.

But as days passed the reports got better. They seemed to indicate that the "flares" were getting larger and more people were reporting seeing them. It was doubtful if this "growth" was psychological because there had been no publicity—so the Air Force decided to reconsider the "flare" answer. They were in the process of doing this on the night of December 5, 1948, a memorable night in the green fireball chapter of UFO history.

At 9:27 P.M. on December 5, an Air Force C-47 transport was flying at 18,000 feet 10 miles east of Albuquerque. The pilot was a Captain Goede. Suddenly the crew, Captain Goede, his co-pilot, and his engineer were startled by a green ball of fire flashing across the sky ahead of them. It looked something like a huge meteor except that it was a bright green color and it didn't arch downward, as meteors usually do. The green-colored ball of fire had started low, from near the eastern slopes of the Sandia Mountains, arched upward a little, then seemed to level out. And it was too big for a meteor, at least it was larger than any meteor that anyone in the C-47 had ever seen before. After a hasty discussion the crew decided that they'd better tell somebody about it, especially since they had seen an identical object twenty-two minutes before near Las Vegas, New Mexico.

Captain Goede picked up his microphone and called the control tower at Kirtland AFB and reported what he and his crew had seen. The tower relayed the message to the local intelligence people.

A few minutes later the captain of Pioneer Airlines Flight 63 called Kirtland Tower. At 9:35 P.M. he had also seen a green ball of fire just east of Las Vegas, New Mexico. He was on his way to Albuquerque and would make a full report when he landed.

When he taxied his DC-3 up to the passenger ramp at Kirtland a few minutes later, several intelligence officers were waiting for him. He reported that at 9:35 P.M. he was on a westerly heading, approaching Las Vegas from the east, when he and his co-pilot saw what they first thought was a "shooting star." It was ahead and a little above them. But, the captain said, it took them only a split second to realize that whatever they saw was too low and had too flat a trajectory to be a meteor. As they watched, the object seemed to approach their airplane head on, changing color from orange red to green. As it became bigger and bigger, the captain said, he thought sure it was going to collide with them so he racked the DC-3 up in a tight turn. As the green ball of fire got abreast of them it began to fall toward the ground, getting dimmer and dimmer until it disappeared. Just before he swerved the DC-3, the fireball was as big, or bigger, than a full moon.

The intelligence officers asked a few more questions and went back to their office. More reports, which had been phoned in from all over northern

New Mexico, were waiting for them. By morning a full-fledged investigation was under way.

No matter what these green fireballs were, the military was getting a little edgy. They might be common meteorites, psychologically enlarged flares, or true UFO's, but whatever they were they were playing around in one of the most sensitive security areas in the United States. Within 100 miles of Albuquerque were two installations that were the backbone of the atomic bomb program, Los Alamos and Sandia Base. Scattered throughout the countryside were other installations vital to the defense of the U.S.: radar stations, fighter-interceptor bases, and the other mysterious areas that had been blocked off by high chain-link fences.

Since the green fireballs bore some resemblance to meteors or meteorites, the Kirtland intelligence officers called in Dr. Lincoln La Paz.

Dr. La Paz said that he would be glad to help, so the officers explained the strange series of events to him. True, he said, the description of the fireballs did sound as if they might be meteorites —except for a few points. One way to be sure was to try to plot the flight path of the green fireballs the same way he had so successfully plotted the flight path of meteorites in the past. From this flight path he could determine where they would have hit the earth—if they were meteorites. They would search this area, and if they found parts of a meteorite they would have the answer to the green fireball riddle.

The fireball activity on the night of December 5 was made to order for plotting flight paths. The good reports of that night included carefully noted locations, the directions in which the green objects were seen, their heights above the horizon, and the times when they were observed. So early the next morning Dr. La Paz and a crew of intelligence officers were scouring northern New Mexico. They started out by talking to the people who had made reports but soon found out that dozens of other people had also seen the fireballs. By closely checking the time of the observations, they determined that eight separate fireballs had been seen. One was evidently more spectacular and was seen by the most people. Everyone in northern New Mexico had seen it going from west to east, so Dr. La Paz and his crew worked eastward across New Mexico to the west border of Texas, talking to dozens of people. After many sleepless hours they finally plotted where it should have struck the earth. They searched the area but found nothing. They went back over the area time and time again— nothing. As Dr. La Paz later told me, this was the first time that he seriously doubted the green fireballs were meteorites.

Within a few more days the fireballs were appearing almost nightly. The intelligence officers from Kirtland decided that maybe they could get a good look at one of them, so on the night of December 8 two officers took off in an airplane just before dark and began to cruise around north of Albuquerque. They had a carefully worked out plan where each man would observe certain details if they saw one of the green fireballs. At 6:33P.M. they saw one. This is their report:

At 6:33P.M. while flying at an indicated altitude of 11,500 feet, a strange phenomenon was observed. Exact position of the aircraft at time of the observation was 20 miles east of the Las Vegas, N.M., radio range station. The aircraft was on a compass course of 90 degrees. Capt. ——— was pilot and I was acting as copilot. I first observed the object and a split second later the pilot saw it. It was 2,000 feet higher than the plane, and was approaching the plane at a rapid rate of speed from 30 degrees to the left of our course. The object was similar in appearance to a burning green flare, the kind that is commonly used in the Air Force. However, the light was much more intense and the object appeared considerably larger than a normal flare. The trajectory of the object, when first sighted, was almost flat and parallel to the earth. The phenomenon lasted about 2 seconds. At the end of this time the object seemed to begin to burn out and the trajectory then dropped off rapidly. The phenomenon was of such intensity as to be visible from the very moment it ignited.

Back at Wright-Patterson AFB, ATIC was getting a blow-by-blow account of the fireball activity but they were taking no direct part in the investigation. Their main interest was to review all incoming UFO reports and see if the green fireball reports were actually unique to the Albuquerque area. They were. Although a good many UFO reports were coming in from other parts of the U.S., none fit the description of the green fireballs.

All during December 1948 and January 1949 the green fireballs continued to invade the New Mexico skies. Everyone, including the intelligence officers at Kirtland AFB, Air Defense Command people, Dr. La Paz, and some of the most distinguished scientists at Los Alamos had seen at least one.

In mid-February 1949 a conference was called at Los Alamos to determine what should be done to further pursue the investigation. The Air Force, Project Sign, the intelligence people at Kirtland, and other interested parties had done everything they could think of and still no answer.

Such notable scientists as Dr. Joseph Kaplan, a world-renowned authority on the physics of the upper atmosphere, Dr. Edward Teller, of H-bomb fame, and of course Dr. La Paz, attended, along with a lot of military brass and scientists from Los Alamos.

This was one conference where there was no need to discuss whether or not this special type of UFO, the green fireball, existed. Almost everyone at the meeting had seen one. The purpose of the conference was to decide whether the fireballs were natural or man-made and how to find out more about them.

As happens in any conference, opinions were divided. Some people thought the green fireballs were natural fireballs. The proponents of the natural meteor, or meteorite, theory presented facts that they had dug out of astronomical journals. Greenish-colored meteors, although not common, had been observed on many occasions. The flat trajectory, which seemed to be so important in proving that the green fireballs were extraterrestrial, was also nothing new. When viewed from certain angles, a meteor can appear to have a flat trajectory. The reason that so many had been seen during December of 1948 and January of 1949 was that the weather had been unusually clear all over the Southwest during this period.

Dr. La Paz led the group who believed that the green fireballs were not meteors or meteorites. His argument was derived from the facts that he had gained after many days of research and working with Air Force intelligence teams. He stuck to the points that (1) the trajectory was too flat, (2) the color was too green, and (3) he couldn't locate any fragments even though he had found the spots where they should have hit the earth if they were meteorites.

People who were at that meeting have told me that Dr. La Paz's theory was very interesting and that each point was carefully considered. But evidently it wasn't conclusive enough because when the conference broke up, after two days, it was decided that the green fireballs were a natural phenomenon of some kind. It was recommended that this phase of the UFO investigation be given to the Air Force's Cambridge Research Laboratory, since it is the function of this group to study natural phenomena, and that Cambridge set up a project to attempt to photograph the green fireballs and measure their speed, altitude, and size.

In the late summer of 1949, Cambridge established Project Twinkle to solve the mystery. The project called for establishing three cinetheodolite stations near White Sands, New Mexico. A cinetheodolite is similar to a 35-mm. movie camera except when you take a photograph of an object you also get a photograph of three dials that show the time the photo was taken, the azimuth angle, and the elevation angle of the camera. If two or more cameras photograph the same object, it is possible to obtain a very accurate measurement of the photographed object's altitude, speed, and size.

Project Twinkle was a bust. Absolutely nothing was photographed. Of the three cameras that were planned for the project, only one was available. This one camera was continually being moved from place to place. If several reports came from a certain area, the camera crew would load up their equipment and move to that area, always arriving too late. Any duck hunter can tell you that this is the wrong tactic; if you want to shoot any ducks pick a good place and stay put, let the ducks come to you.

The people trying to operate Project Twinkle were having financial and morale trouble. To do a good job they needed more and better equipment and more people, but Air Force budget cuts precluded this. Moral support was free but they didn't get this either.

When the Korean War started, Project Twinkle silently died, along with official interest in green fireballs.

When I organized Project Blue Book in the summer of 1951 I'd never heard of a green fireball. We had a few files marked "Los Alamos Conference," "Fireballs," "Project Twinkle," etc., but I didn't pay any attention to them.

Then one day I was at a meeting in Los Angeles with several other officers from ATIC, and was introduced to Dr. Joseph Kaplan. When he found we were from ATIC, his first question was, "What ever happened to the green fireballs?" None of us had ever heard of them, so he quickly gave us the story. He and I ended up discussing green fireballs. He mentioned Dr. La Paz and his opinion that the green fireballs might be man-made, and although he respected La Paz's professional ability, he just wasn't convinced. But he did strongly urge me to get in touch with Dr. La Paz and hear his side of the story.

When I returned to ATIC I spent several days digging into our collection of green fireball reports. All of these reports covered a period from early December 1948 to 1949. As far as Blue Book's files were concerned, there hadn't been a green fireball report for a year and a half.

I read over the report on Project Twinkle and the few notes we had on the Los Alamos Conference, and decided that the next time I went to Albuquerque I'd contact Dr. La Paz. I did go to Albuquerque several times but my visits were always short and I was always in a hurry so I didn't get to see him.

It was six or eight months later before the subject of green fireballs came up again. I was eating lunch with a group of people at the AEC's Los Alamos Laboratory when one of the group mentioned the mysterious kelly-green balls of fire. The strictly unofficial bull- session-type discussion that followed took up the entire lunch hour and several hours of the afternoon. It was an interesting discussion because these people, all scientists and technicians from the lab, had a few educated guesses as to what they might be. All of them had seen a green fireball, some of them had seen several.

One of the men, a private pilot, had encountered a fireball one night while he was flying his Navion north of Santa Fe and he had a vivid way of explaining what he'd seen. "Take a soft ball and paint it with some kind of fluorescent paint that will glow a bright green in the dark," I remember his saying, "then have someone take the ball out about 100 feet in front of you and about 10 feet above you. Have him throw the ball right at your face, as hard as he can throw it. That's what a green fireball looks like."

The speculation about what the green fireballs were ran through the usual spectrum of answers, a new type of natural phenomenon, a secret U.S. development, and psychologically enlarged meteors. When the possibility of the green fireballs' being associated with interplanetary vehicles came up, the whole group got serious. They had been doing a lot of thinking about this, they said, and they had a theory.

The green fireballs, they theorized, could be some type of unmanned test vehicle that was being projected into our atmosphere from a "spaceship" hovering several hundred miles above the earth. Two years ago I would have been amazed to hear a group of reputable scientists make such a startling statement. Now, however, I took it as a matter of course. I'd heard the same type of statement many times before from equally qualified groups.

Turn the tables, they said, suppose that we are going to try to go to a far planet. There would be three phases to the trip: out through the earth's atmosphere, through space, and the re-entry into the atmosphere of the planet we're planning to land on. The first two phases would admittedly present formidable problems, but the last phase, the re-entry phase, would be the most critical. Coming in from outer space, the craft would, for all practical purposes, be similar to a meteorite except that it would be powered and not free-falling. You would have myriad problems associated with aerodynamic heating, high aerodynamic loadings, and very probably a host of other problems that no one can now conceive of. Certain of these problems could be partially solved by laboratory experimentation, but nothing can replace flight testing, and the results obtained by flight tests in our atmosphere would not be valid in another type of atmosphere. The most logical way to overcome this difficulty would be to build our interplanetary vehicle, go to the planet that we were interested in landing on, and hover several hundred miles up. From this altitude we could send instrumented test vehicles down to the planet. If we didn't want the inhabitants of the planet, if it were inhabited, to know what we were doing we could put destruction devices in the test vehicle, or arrange the test so that the test vehicles would just plain burn up at a certain point due to aerodynamic heating.

They continued, each man injecting his ideas.

Maybe the green fireballs are test vehicles—somebody else's. The regular UFO reports might be explained by the fact that the manned vehicles were venturing down to within 100,000 or 200,000 feet of the earth, or to the altitude at which atmosphere re-entry begins to get critical.

I had to go down to the airstrip to get a CARCO Airlines plane back to Albuquerque so I didn't have time to ask a lot of questions that came into my mind. I did get to make one comment. From the conversations, I assumed that these people didn't think the green fireballs were any kind of a natural phenomenon. Not exactly, they said, but so far the evidence that said they were a natural phenomenon was vastly outweighed by the evidence that said they weren't.

During the kidney-jolting trip down the valley from Los Alamos to Albuquerque in one of the CARCO Airlines' Bonanzas, I decided that I'd stay over an extra day and talk to Dr. La Paz.

He knew every detail there was to know about the green fireballs. He confirmed my findings, that the genuine green fireballs were no longer being seen. He said that he'd received hundreds of reports, especially after he'd written several articles about the mysterious fireballs, but that all of the reported objects were just greenish- colored, common, everyday meteors.

Dr. La Paz said that some people, including Dr. Joseph Kaplan and Dr. Edward Teller, thought that the green fireballs were natural meteors. He didn't think so, however, for several reasons. First the color was so much different. To illustrate his point, Dr. La Paz opened his desk drawer and took out a well-worn chart of the color spectrum. He checked off two shades of green; one a pale, almost yellowish green and the other a much more distinct vivid green. He pointed to the bright green and told me that this was the color of the green fireballs. He'd taken this chart with him when he went out to talk to people who had seen the green fireballs and everyone had picked this one color. The pale green, he explained, was the color reported in the cases of documented green meteors.

Then there were other points of dissimilarity between a meteor and the green fireballs. The trajectory of the fireballs was too flat. Dr. La Paz explained that a meteor doesn't necessarily have to arch down across the sky, its trajectory can appear to be flat, but not as flat as that of the green fireballs. Then there was the size. Almost always such descriptive words as "terrifying," "as big as the moon," and "blinding" had been used to describe the fireballs. Meteors just aren't this big and bright.

No—Dr. La Paz didn't think that they were meteors.

Dr. La Paz didn't believe that they were meteorites either.

A meteorite is accompanied by sound and shock waves that break windows and stampede cattle. Yet in every case of a green fireball sighting the observers reported that they did not hear any sound.

But the biggest mystery of all was the fact that no particles of a green fireball had ever been found. If they were meteorites, Dr. La Paz was positive that he would have found one. He'd missed very few times in the cases of known meteorites. He pulled a map out of his file to show me what he meant. It was a map that he had used to plot the spot where a meteorite had hit the earth. I believe it was in Kansas. The map had been prepared from information he had obtained from dozens of people who had seen the meteorite come flaming toward the earth. At each spot where an observer was standing he'd drawn in the observer's line of sight to the meteorite. From the dozens of observers he had obtained dozens of lines of sight. The lines all converged to give Dr. La Paz a plot of the meteorite's downward trajectory. Then he had been able to plot the spot where it had struck the earth. He and his crew went to the marked area, probed the ground with long steel poles, and found the meteorite.

This was just one case that he showed me. He had records of many more similar successful expeditions in his file.

Then he showed me some other maps. The plotted lines looked identical to the ones on the map I'd just seen. Dr. La Paz had used the same techniques on these plots and had marked an area where he wanted to search. He had searched the area many times but he had never found anything.

These were plots of the path of a green fireball.

When Dr. La Paz had finished, I had one last question, "What do you think they are?"

He weighed the question for a few seconds—then he said that all he cared to say was that he didn't think that they were a natural phenomenon. He thought that maybe someday one would hit the earth and the mystery would be solved. He hoped that they were a natural phenomenon.

After my talk with Dr. La Paz I can well understand his apparent calmness on the night of September 18, 1954, when the newspaper reporter called him to find out if he planned to investigate this latest green fireball report. He was speaking from experience, not indifference, when he said, "But I don't expect to find anything."

If the green fireballs are back, I hope that Dr. La Paz gets an answer this time.

The story of the UFO now goes back to late January 1949, the time when the Air Force was in the midst of the green fireball mystery. In another part of the country another odd series of events was taking place. The center of activity was a highly secret area that can't be named, and the recipient of the UFO's, which were formations of little lights, was the U.S. Army.

The series of incidents started when military patrols who were protecting the area began to report seeing formations of lights flying through the night sky. At first the lights were reported every three or four nights, but inside of two weeks the frequency had stepped up. Before long they were a nightly occurrence. Some patrols reported that they had seen three or four formations in one night. The sightings weren't restricted to the men on patrol. One night, just at dusk, during retreat, the entire garrison watched a formation pass directly over the post parade ground.

As usual with UFO reports, the descriptions of the lights varied but the majority of the observers reported a V formation of three lights. As the formation moved through the sky, the lights changed in color from a bluish white to orange and back to bluish white. This color cycle took about two seconds. The lights usually traveled from west to east and made no sound. They didn't streak across the sky like a meteor, but they were "going faster than a jet." The lights were "a little bigger than the biggest star." Once in a while the GI's would get binoculars on them but they couldn't see any more details. The lights just looked bigger.

From the time of the first sighting, reports of the little lights were being sent to the Air Force through Army Intelligence channels. The reports were getting to ATIC, but the green fireball activity was taking top billing and no comments went back to the Army about their little lights. According to an Army G-2 major to whom I talked in the Pentagon, this silence was taken to mean that no action, other than sending in reports, was necessary on the part of the Army.

But after about two weeks of nightly sightings and no apparent action by the Air Force, the commander of the installation decided to take the initiative and set a trap. His staff worked out a plan in record time. Special UFO patrols would be sent out into the security area and they would be furnished with sighting equipment. This could be the equipment that they normally used for fire control. Each patrol would be sent to a specific location and would set up a command post. Operating out of the command post, at points where the sky could be observed, would be sighting teams. Each team had sighting equipment to measure the elevation and azimuth angle of the UFO. Four men were to be on each team, an instrument man, a timer, a recorder, and a radio operator. All the UFO patrols would be assigned special radio frequencies.

54

The operating procedure would be that when one sighting team spotted a UFO the radio operator would call out his team's location, the location of the UFO in the sky, and the direction it was going. All of the other teams from his patrol would thus know when to look for the UFO and begin to sight on it. While the radio man was reporting, the instrument man on the team would line up the UFO and begin to call out the angles of elevation and azimuth. The timer would call out the time; the recorder would write all of this down. The command post, upon hearing the report of the UFO, would call the next patrol and tell them. They too would try to pick it up.

Here was an excellent opportunity to get some concrete data on at least one type of UFO. It was something that should have been done from the start. Speeds, altitudes, and sizes that are estimated just by looking at a UFO are miserably inaccurate. But if you could accurately establish that some type of object was traveling 30,000 miles an hour—or even 3,000 miles an hour—through our atmosphere, the UFO story would be the biggest story since the Creation.

The plan seemed foolproof and had the full support of every man who was to participate. For the first time in history every GI wanted to get on the patrols. The plan was quickly written up as a field order, approved, and mimeographed. Since the Air Force had the prime responsibility for the UFO investigation, it was decided that the plan should be quickly co-ordinated with the Air Force, so a copy was rushed to them. Time was critical because every group of nightly reports might be the last. Everything was ready to roll the minute the Air Force said "Go."

The Air Force didn't O.K. the plan. I don't know where the plan was killed, or who killed it, but it was killed. Its death caused two reactions.

Many people thought that the plan was killed so that too many people wouldn't find out the truth about UFO's. Others thought somebody was just plain stupid. Neither was true. The answer was simply that the official attitude toward UFO's had drastically changed in the past few months. They didn't exist, they couldn't exist. It was the belief at ATIC that the one last mystery, the green fireballs, had been solved a few days before at Los Alamos. The fireballs were meteors and Project Twinkle would prove it. Any further investigation by the Army would be a waste of time and effort.

This drastic change in official attitude is as difficult to explain as it was difficult for many people who knew what was going on inside Project Sign to believe. I use the words "official attitude" because at this time UFO's had become as controversial a subject as they are today. All through intelligence circles people had chosen sides and the two UFO factions that exist today were born.

On one side was the faction that still believed in flying saucers. These people, come hell or high water, were hanging on to their original ideas. Some thought that the UFO's were interplanetary spaceships. Others weren't quite as bold and just believed that a good deal more should be known about the UFO's before they were so completely written off. These people weren't a bunch of nuts or crackpots either. They ranged down through the ranks from generals and top-grade civilians. On the outside their views were backed up by civilian scientists.

On the other side were those who didn't believe in flying saucers. At one time many of them had been believers. When the UFO reports were pouring in back in 1947 and 1948, they were just as sure that the UFO's were real as the people they were now scoffing at. But they had changed their minds. Some of them had changed their minds because they had seriously studied the UFO reports and just couldn't see any evidence that the UFO's were real. But many of them could see the "I don't believe" band wagon pulling out in front and just jumped on.

This change in the operating policy of the UFO project was so pronounced that I, like so many other people, wondered if there was a hidden reason for the change. Was it actually an attempt to go underground—to make the project more secretive? Was it an effort to cover up the fact that UFO's were proven to be interplanetary and that this should be withheld from the public at all cost to prevent a mass panic? The UFO files are full of references to the near mass panic of October 30, 1938, when Orson Welles presented his now famous "The War of the Worlds" broadcast.

This period of "mind changing" bothered me. Here were people deciding that there was nothing to this UFO business right at a time when the reports seemed to be getting better. From what I could see, if there was any mind changing to be done it should have been the other way, skeptics should have been changing to believers.

Maybe I was just playing the front man to a big cover-up. I didn't like it because if somebody up above me knew that UFO's were really spacecraft, I could make a big fool out of myself if the truth came out. I checked into this thoroughly. I spent a lot of time talking to people who had worked on Project Grudge.

The anti-saucer faction was born because of an old psychological trait, people don't like to be losers. To be a loser makes one feel inferior and incompetent. On September 23, 1947, when the chief of ATIC sent a letter to the Commanding General of the Army Air Forces stating that UFO's were real, intelligence committed themselves. They had to prove it. They tried for a year and a half with no success. Officers on top began to get anxious and the press began to get anxious. They wanted an answer. Intelligence had tried one answer, the then Top Secret Estimate of the Situation that "proved" that UFO's were real, but it was kicked back. The people on the UFO project began to think maybe the brass didn't consider them too sharp so they tried a new hypothesis: UFO's don't exist. In no time they found that this was easier to prove and it got recognition. Before if an especially interesting UFO report came in and the Pentagon wanted an answer, all they'd get was an "It could be real but we can't prove it." Now such a request got a quick, snappy "It was a balloon," and feathers were stuck in caps from ATIC up to the Pentagon. Everybody felt fine.

In early 1949 the term "new look" was well known. The new look in women's fashions was the lower hemlines, in automobiles it was longer lines. In UFO circles the new look was cuss 'em.

The new look in UFO's was officially acknowledged on February 11, 1949, when an order was written that changed the name of the UFO project from Project Sign to Project Grudge. The order was supposedly written because the classified name, Project Sign, had been compromised. This was always my official answer to any questions about the name change. I'd go further and say that the names of the projects, first Sign, then Grudge, had no significance. This wasn't true, they did have significance, a lot of it.

Chapter Five - The Dark Ages

The order of February 11, 1949, that changed the name of Project Sign to Project Grudge had not directed any change in the operating policy of the project. It had, in fact, pointed out that the project was to continue to investigate and evaluate reports of sightings of unidentified flying objects. In doing this, standard intelligence procedures would be used. This normally means the *unbiased evaluation* of intelligence data. But it doesn't take a great deal of study of the old UFO files to see that standard intelligence procedures were no longer being used by Project Grudge. Everything was being evaluated on the premise that UFO's couldn't exist. No matter what you see or hear, don't believe it.

New people took over Project Grudge. ATIC's top intelligence specialists who had been so eager to work on Project Sign were no longer working on Project Grudge. Some of them had drastically and hurriedly changed their minds about UFO's when they thought that the Pentagon was no longer sympathetic to the UFO cause. They were now directing their talents toward more socially acceptable projects. Other charter members of Project Sign had been "purged." These were the people who had refused to change their original opinions about UFO's.

With the new name and the new personnel came the new objective, get rid of the UFO's. It was never specified this way in writing but it didn't take much effort to see that this was the goal of Project Grudge. This unwritten objective was reflected in every memo, report, and directive.

To reach their objective Project Grudge launched into a campaign that opened a new age in the history of the UFO. If a comparative age in world history can be chosen, the Dark Ages would be most appropriate. Webster's Dictionary defines the Dark Ages as a period of "intellectual stagnation."

To one who is intimately familiar with UFO history it is clear that Project Grudge had a two-phase program of UFO annihilation. The first phase consisted of explaining every UFO report. The second phase was to tell the public how the Air Force had solved all the sightings. This, Project Grudge reasoned, would put an end to UFO reports.

Phase one had been started by the people of Project Sign. They realized that a great many reports were caused by people seeing balloons or such astronomical bodies as planets, meteors, or stars. They also realized that before they could get to the heart of the UFO problems they had to sift out this type of report. To do this they had called on outside help. Air Weather Service had been asked to screen the reports and check those that sounded like balloons against their records of balloon flights. Dr. J. Allen Hynek, distinguished astrophysicist and head of Ohio State University's Astronomy Department, had been given a contract to sort out those reports that could be blamed on stars, planets, meteors, etc. By early March the Air Weather Service and Dr. Hynek had some positive identifications. According to the old records, with these solutions and those that Sign and Grudge had already found, about 50 per cent of the reported UFO's could now be positively identified as hoaxes, balloons, planets, sundogs, etc. It was now time to start phase two, the publicity campaign.

For many months reporters and writers had been trying to reach behind the security wall and get the UFO story from the horse's mouth, but no luck. Some of them were still trying but they were having no success because they were making the mistake of letting it slip that they didn't believe that airline pilots, military pilots, scientists, and just all around solid citizens were having "hallucinations," perpetrating "hoaxes," or being deceived by the "misidentification of common objects." The people of Project Grudge weren't looking for this type of writer, they wanted a writer who would listen to them and write their story. As a public relations officer later told me, "We had a devil of a time. All of the writers who were after saucer stories had made their own investigations of sightings and we couldn't convince them they were wrong."

Before long, however, the right man came along. He was Sidney Shallet, a writer for *The Saturday Evening Post*. He seemed to have the prerequisites that were desired, so his visit to ATIC was cleared through the Pentagon. Harry Haberer, a crack Air Force public relations man, was assigned the job of seeing that Shallet got his story. I have heard many times, from both military personnel and civilians, that the Air Force told Shallet exactly what to say in his article—play down the UFO's—don't write anything that even hints that there might be something foreign in our skies. I don't believe that this is the case. I think that he just wrote the UFO story as it was told to him, told to him by Project Grudge.

Shallet's article, which appeared in two parts in the April 30 and May 7, 1949, issues of *The Saturday Evening Post*, is important in the history of the UFO and in understanding the UFO problem because it had considerable effect on public opinion. Many people had, with varying degrees of interest, been wondering about the UFO's for over a year and a half. Very few had any definite opinions one way or the other. The feeling seemed to be that the Air

Force is working on the problem and when they get the answer we'll know. There had been a few brief, ambiguous press releases from the Air Force but these meant nothing. Consequently when Shallet's article appeared in the *Post* it was widely read. It contained facts, and the facts had come from Air Force Intelligence. This was the Air Force officially reporting on UFO's for the first time.

The article was typical of the many flying saucer stories that were to follow in the later years of UFO history, all written from material obtained from the Air Force. Shallet's article casually admitted that a few UFO sightings couldn't be explained, but the reader didn't have much chance to think about this fact because 99 per cent of the story was devoted to the anti-saucer side of the problem. It was the typical negative approach. I know that the negative approach is typical of the way that material is handed out by the Air Force because I was continually being told to "tell them about the sighting reports we've solved—don't mention the unknowns." I was never ordered to tell this, but it was a strong suggestion and in the military when higher headquarters suggests, you do.

Shallet's article started out by psychologically conditioning the reader by using such phrases as "the great flying saucer scare," "rich, full-blown screwiness," "fearsome freaks," and so forth. By the time the reader gets to the meat of the article he feels like a rich, full-blown jerk for ever even thinking about UFO's.

He pointed out how the "furor" about UFO reports got so great that the Air Force was "forced" to investigate the reports reluctantly. He didn't mention that two months after the first UFO report ATIC had asked for Project Sign since they believed that UFO's did exist. Nor did it mention the once Top Secret Estimate of the Situation that also concluded that UFO's were real. In no way did the article reflect the excitement and anxiety of the age of Project Sign when secret conferences preceded and followed every trip to investigate a UFO report. This was the Air Force being "forced" into reluctantly investigating the UFO reports.

Laced through the story were the details of several UFO sightings; some new and some old, as far as the public was concerned. The original UFO report by Kenneth Arnold couldn't be explained. Arnold, however, had sold his story to *Fate* magazine and in the same issue of *Fate* were stories with such titles as "Behind the Etheric Veil" and "Invisible Beings Walk the Earth," suggesting that Arnold's story might fall into the same category. The sightings where the Air Force had the answer had detailed explanations. The ones that were unknowns were mentioned, but only in passing.

Many famous names were quoted. The late General Hoyt S. Vanden-berg, then Chief of Staff of the Air Force, had seen a flying saucer but it was just a reflection on the windshield of his B-17. General Lauris Norstad's UFO was a reflection of a star on a cloud, and General Curtis E. Le May found out that one out of six UFO's was a balloon; Colonel McCoy, then chief of ATIC, had seen lots of UFO's. All were reflections from distant airplanes. In other words, nobody who is anybody in the Air Force believes in flying saucers.

Figures in the top echelons of the military had spoken.

A few hoaxes and crackpot reports rounded out Mr. Shallet's article.

The reaction to the article wasn't what the Air Force and ATIC expected. They had thought that the public would read the article and toss it, and all thoughts of UFO's, into the trash can. But they didn't. Within a few days the frequency of UFO reports hit an all-time high. People, both military and civilian, evidently didn't much care what Generals Vandenberg, Norstad, Le May, or Colonel McCoy thought; they didn't believe what they were seeing were hallucinations, reflections, or balloons. What they were seeing were UFO's, whatever UFO's might be.

I heard many times from ex-Project Grudge people that Shallet had "crossed" them, he'd vaguely mentioned that there might be a case for the UFO. This made him pro-saucer.

A few days after the last installment of the *Post* article the Air Force gave out a long and detailed press release completely debunking UFO's, but this had no effect. It only seemed to add to the confusion.

The one thing that Shallet's article accomplished was to plant a seed of doubt in many people's minds. Was the Air Force telling the truth about UFO's? The public and a large percentage of the military didn't know what was going on behind ATIC's barbed-wire fence but they did know that a lot of reliable people had seen UFO's. Airline pilots are considered responsible people—airline pilots had seen UFO's. Experienced military pilots and ground officers are responsible people—they'd seen UFO's. Scientists, doctors, lawyers, merchants, and plain old Joe Doakes had seen UFO's, and their friends knew that they were responsible people. Somehow these facts and the tone of the *Post* article didn't quite jibe, and when things don't jibe, people get suspicious.

In those people who had a good idea of what was going on behind ATIC's barbed wire, the newspaper reporters and writers with the "usually reliable sources," the *Post* article planted a bigger seed of doubt. Why the sudden change in policy they wondered? If UFO's were so serious a few months ago, why the sudden debunking? Maybe Shallet's story was a put-up job for the Air Force. Maybe the security had been tightened. Their sources of information were reporting that many people in the military did not quite buy the Shallet article. The seed of doubt began to grow, and some of these writers began to start "independent investigations" to get the "true" story. Research takes time, so during the summer and fall of 1949 there wasn't much apparent UFO activity.

As the writers began to poke around for their own facts, Project Grudge lapsed more and more into a period of almost complete inactivity. Good UFO reports continued to come in at the rate of about ten per month but they weren't being verified or investigated. Most of them were being discarded. There are few, if any, UFO reports for the middle and latter part of 1949 in the ATIC files. Only the logbook, showing incoming reports, gives any idea of the activity of this period. The meager effort that was being made was going into a report that evaluated old UFO reports, those received prior to the spring of 1949. Project Grudge *thought* that they were writing a final report on the UFO's.

From the small bits of correspondence and memos that were in the ATIC files, it was apparent that Project Grudge thought that the UFO was on its way out. Any writers inquiring about UFO activity were referred to the debunking press release given out just after the *Post* article had been published. There was no more to say. Project Grudge thought they were winning the UFO battle; the writers thought that they were covering up a terrific news story—the story that the Air Force knew what flying saucers were and weren't telling.

By late fall 1949 the material for several UFO stories had been collected by writers who had been traveling all over the United States talking to people who had seen UFO's. By early winter the material had been worked up into UFO stories. In December the presses began to roll. *True* magazine "scooped" the world with their story that UFO's were from outer space.

The *True* article, entitled, "The Flying Saucers Are Real," was written by Donald Keyhoe. The article opened with a hard punch. In the first paragraph Keyhoe concluded that after eight months of extensive research he had found evidence that the earth was being closely scrutinized by intelligent beings. Their vehicles were the so- called flying saucers. Then he proceeded to prove his point. His argument was built around the three classics: the Mantell, the Chiles- Whitted, and the Gorman incidents. He took each sighting, detailed the "facts," ripped the official Air Force conclusions to shreds, and presented his own analysis. He threw in a varied assortment of technical facts that gave the article a distinct, authoritative flavor. This, combined with the fact that *True* had the name for printing the truth, hit the reading public like an 8-inch howitzer. Hours after it appeared in subscribers' mailboxes and on the newsstands, radio and TV commentators and newspapers were giving it a big play. UFO's were back in business, to stay. True was in business too. It is rumored among magazine publishers that Don Keyhoe's article in *True* was one of the most widely read and widely discussed magazine articles in history.

The Air Force had inadvertently helped Keyhoe—in fact, they made his story a success. He and several other writers had contacted the Air Force asking for information for their magazine articles. But, knowing that the articles were pro-saucer, the writers were unceremoniously sloughed off. Keyhoe carried his fight right to the top, to General Sory Smith, Director of the Office of Public Information, but still no dice—the Air Force wasn't divulging any more than they had already told. Keyhoe construed this to mean tight security, the tightest type of security. Keyhoe had one more approach, however. He was an ex-Annapolis graduate, and among his classmates were such people as Admiral Delmar Fahrney, then a top figure in the Navy guided missile program and Admiral Calvin Bolster, the Director of the Office of Naval

Research. He went to see them but they couldn't help him. He *knew* that this meant the real UFO story was big and that it could be only one thing— interplanetary spaceships or earthly weapons—and his contacts denied they were earthly weapons. He played this security angle in his *True* article and in a later book, and it gave the story the needed punch.

But the Air Force wasn't trying to cover up. It was just that they didn't want Keyhoe or any other saucer fans in their hair. They couldn't be bothered. They didn't believe in flying saucers and couldn't feature anybody else believing. Believing, to the people in ATIC in 1949, meant even raising the possibility that there might be something to the reports.

The Air Force had a plan to counter the Keyhoe article, or any other story that might appear. The plan originated at ATIC. It called for a general officer to hold a short press conference, flash his stars, and speak the magic words "hoaxes, hallucinations, and the misidentification of known objects," *True*, Keyhoe and the rest would go broke trying to peddle their magazines. The *True* article did come out, the general spoke, the public laughed, and Keyhoe and *True* got rich. Only the other magazines that had planned to run UFO stories, and that were scooped by *True*, lost out. Their stories were killed—they would have been an anti-climax to Keyhoe's potboiler.

The Air Force's short press conference was followed by a press release. On December 27, 1949, it was announced that Project Grudge had been closed out and the final report on UFO's would be released to the press in a few days. When it was released it caused widespread interest because, supposedly, this was all that the Air Force knew about UFO's. Once again, instead of throwing large amounts of cold water on the UFO's, it only caused more confusion.

The report was officially titled "Unidentified Flying Objects— Project Grudge," Technical Report No. 102-AC-49/15-100. But it was widely referred to as the Grudge Report.

The Grudge Report was a typical military report. There was the body of the report, which contained the short discussion, conclusions, and recommendations. Then there were several appendixes that were supposed to substantiate the conclusions and recommendations made in the report.

One of the appendixes was the final report of Dr. J. Allen Hynek, Project Grudge's contract astronomer. Dr. Hynek and his staff had studied 237 of the best UFO reports. They had spent several months analyzing each report. By searching through astronomical journals and checking the location of various celestial bodies, they found that some UFO's could be explained. Of the 237 reports he and his staff examined, 32 per cent could be explained astronomically.

The Air Force Air Weather Service and the Air Force Cambridge Research Laboratory had sifted the reports for UFO's that might have been balloons. These two organizations had data on the flights of both the regular weather balloons and the huge, high-flying skyhooks. They wrote off 12 per cent of the 237 UFO reports under study as balloons.

This left 56 per cent still unknown. By weeding out the hoaxes, the reports that were too nebulous to evaluate, and reports that could well be misidentified airplanes, Project Grudge disposed of another 33 per cent of the reports. This left 23 per cent that fell in the "unknown" category.

There were more appendixes. The Rand Corporation, one of the most unpublicized yet highly competent contractors to the Air Force, looked over the reports and made the statement, "We have found nothing which would seriously controvert simple rational explanations of the various phenomena in terms of balloons, conventional aircraft, planets, meteors, bits of paper, optical illusions, practical jokers, psychopathological reporters, and the like." But Rand's comment didn't help a great deal because they didn't come up with any solutions to any of the 23 per cent unknown.

The Psychology Branch of the Air Force's Aeromedical Laboratory took a pass at the psychological angles. They said, "there are sufficient psychological explanations for the reports of unidentified objects to provide plausible explanations for reports not otherwise explainable." They pointed out that some people have "spots in front of their eyes" due to minute solid particles that float about in the fluids of the eye and cast shadows on the retina. Then they pointed out that some people are just plain nuts. Many people who read the Grudge Report took these two points to mean that all UFO observers either had spots in front of their eyes or were nuts. They broke the reports down statistically. The people who wrote the report found that over 70 per cent of the people making sightings reported a light- colored object. (This I doubt, but that's what the report said.) They said a big point of these reports of light-colored objects was that any high-flying object will appear to be dark against the sky. For this reason the UFO's couldn't be real.

I suggest that the next time you are outdoors and see a bomber go over at high altitude you look at it closely. Unless it's painted a dark color it won't look dark.

The U.S. Weather Bureau wrote an extremely comprehensive and interesting report on all types of lightning. It was included in the Grudge Report but contained a note: "None of the recorded incidents appear to have been lightning."

There was one last appendix. It was entitled "Summary of the Evaluation of Remaining Reports." What the title meant was, We have 23 per cent of the reports that we can't explain but we have to explain them because we don't believe in flying saucers. This appendix contributed greatly to the usage of the analogy to the Dark Ages, the age of "intellectual stagnation."

This appendix was important—it was the meat of the whole report. Every UFO sighting had been carefully checked, and those with answers had been sifted out. Then the ones listed in "Summary of the Evaluation of Remaining Reports" should be the best UFO reports—the ones with no answers.

This was the appendix that the newsmen grabbed at when the Grudge Report was released. It contained the big story. But if you'll check back through old newspaper files you will hardly find a mention of the Grudge Report.

I was told that reporters just didn't believe it when I tried to find out why the Grudge Report hadn't been mentioned in the newspapers. I got the story from a newspaper correspondent in Washington whom I came to know pretty well and who kept me filled in on the latest UFO scuttlebutt being passed around the Washington press circles. He was one of those humans who had a brain like a filing cabinet; he could remember everything about everything. UFO's were a hobby of his. He remembered when the Grudge Report came out; in fact, he'd managed to get a copy of his own. He said the report had been quite impressive, but only in its ambiguousness, illogical reasoning, and very apparent effort to write off all UFO reports at any cost. He, personally, thought that it was a poor attempt to put out a "fake" report, full of misleading information, to cover up the real story. Others, he told me, just plainly and simply didn't know what to think—they were confused.

And they had every right to be confused.

As an example of the way that many of the better reports of the 1947- 49 period were "evaluated" let's take the report of a pilot who tangled with a UFO near Washington, D.C., on the night of November 18, 1948.

At about 9:45 EST I noticed a light moving generally north to south over Andrews AFB. It appeared to be one continuous, glowing white light. I thought it was an aircraft with only one landing light so I moved in closer to check, as I wanted to get into the landing pattern. I was well above landing traffic altitude at this time. As I neared the light I noticed that it was not another airplane. Just then it began to take violent evasive action so I tried to close on it. I made first contact at 2,700 feet over the field. I switched my navigation lights on and off but got no answer so I went in closer— but the light quickly flew up and over my airplane. I then tried to close again but the light turned. I tried to turn inside of its turn and, at the same time, get the light between the moon and me, but even with my flaps lowered I couldn't turn inside the light. I never did manage to get into a position where the light was silhouetted against the moon.

I chased the light up and down and around for about 10 minutes, then as a last resort I made a pass and turned on my landing lights. Just before the object made a final tight turn and headed for the coast I saw that it was a dark gray oval-shaped object, smaller than my T-6. I couldn't tell if the light was on the object or if the whole object had been glowing.

Two officers and a crew chief, a master sergeant, completely corroborated the pilot's report. They had been standing on the flight line and had witnessed the entire incident.

The Air Weather Service, who had been called in as experts on weather balloons, read this report. They said, "Definitely not a balloon." Dr. Hynek said, "No astronomical explanation." It wasn't another airplane and it wasn't a hallucination.

But Project Grudge had an answer, it *was* a weather balloon. There was no explanation as to why they had so glibly reversed the decision of the Air Weather Service.

There was an answer for every report.

From the 600 pages of appendixes, discussions of the appendixes, and careful studies of UFO reports, it was concluded that:

Evaluation of reports of unidentified flying objects constitute no direct threat to the national security of the United States.

Reports of unidentified flying objects are the result of:

A mild form of mass hysteria or "war nerves."

Individuals who fabricate such reports to perpetrate a hoax or seek publicity.

Psychopathological persons.

Misidentification of various conventional objects.

It was recommended that Project Grudge be "reduced in scope" and that only "those reports clearly indicating realistic technical applications" be sent to Grudge. There was a note below these recommendations. It said, "It is readily apparent that further study along present lines would only confirm the findings presented herein."

Somebody read the note and concurred because with the completion and approval of the Grudge Report, Project Grudge folded. People could rant and rave, see flying saucers, pink elephants, sea serpents, or Harvey, but it was no concern of ATIC's.

Chapter Six - The Presses Roll—The Air Force Shrugs

The Grudge Report was supposedly not for general distribution. A few copies were sent to the Air Force Press Desk in the Pentagon and reporters and writers could come in and read it. But a good many copies did get into circulation. The Air Force Press Room wasn't the best place to sit and study a 600-page report, and a quick glance at the report showed that it required some study—if no more than to find out what the authors were trying to prove—so several dozen copies got into circulation. I know that these "liberated" copies of the Grudge Report had been thoroughly studied because nearly every writer who came to ATIC during the time that I was in charge of Project Blue Book carried a copy.

Since the press had some questions about the motives behind releasing the Grudge Report, it received very little publicity while the writers put out feelers. Consequently in early 1950 you didn't read much about flying saucers.

Evidently certain people in the Air Force thought this lull in publicity meant that the UFO's had finally died because Project Grudge was junked. All the project files, hundreds of pounds of reports, memos, photos, sketches, and other assorted bits of paper were unceremoniously yanked out of their filing cabinets, tied up with string, and chucked into an old storage case. I would guess that many reports ended up as "souvenirs" because a year later, when I exhumed these files, there were a lot of reports missing.

About this time the official Air Force UFO project had one last post- death muscular spasm. The last bundle of reports had just landed on top of the pile in the storage case when ATIC received a letter from the Director of Intelligence of the Air Force. In official language it said, "What gives?" There had been no order to end Project Grudge. The answer went back that Project Grudge had not been disbanded; the project functions had been transferred and it was no longer a "special" project. From now on UFO reports would be processed through normal intelligence channels along with other intelligence reports.

To show good faith ATIC requested permission to issue a new Air Force-wide bulletin which was duly mimeographed and disseminated. In essence it said that Air Force Headquarters had directed ATIC to continue to collect and evaluate reports of unidentified flying objects. It went on to explain that most UFO reports were trash. It pointed out the findings of the Grudge Report in such strong language that by the time the recipient of the bulletin had finished reading it, he would be ashamed to send in a report. To cinch the deal the bulletins must have been disseminated only to troops in Outer Mongolia because I never found anyone in the field who had ever received a copy.

As the Air Force UFO-investigating activity dropped to nil, the press activity skyrocketed to a new peak. A dozen people took off to dig up their own UFO stories and to draw their own conclusions.

After a quiet January, *True* again clobbered the reading public. This time it was a story in the March 1950 issue and it was entitled, "How Scientists Tracked Flying Saucers." It was written by none other than the man who was at that time in charge of a team of Navy scientists at the super hush-hush guided missile test and development area, White Sands Proving Ground, New Mexico. He was Commander R. B. McLaughlin, an Annapolis graduate and a Regular Navy officer. His story had been cleared by the military and was in absolute, 180- degree, direct contradiction to every press release that had been made by the military in the past two years. Not only did the commander believe that he had proved that UFO's were real but that he knew what they were. "I am convinced," he wrote in the *True* article, "that it," referring to a UFO he had seen at White Sands, "was a flying saucer, and further, that these disks are spaceships from another planet, operated by animate, intelligent beings."

On several occasions during 1948 and 1949, McLaughlin or his crew at the White Sands Proving Ground had made good UFO sightings. The best one was made on April 24, 1949, when the commander's crew of engineers, scientists, and technicians were getting ready to launch one of the huge 100-foot-diameter skyhook balloons. It was 10:30A.M. on an absolutely clear Sunday morning. Prior to the launching, the crew had sent up a small weather balloon to check the winds at lower levels. One man was watching the balloon through a theodolite, an instrument similar to a surveyor's transit built around a 25-power telescope, one man was holding a stop watch, and a third had a clipboard to record the measured data. The crew had tracked the balloon to about 10,000 feet when one of them suddenly shouted and pointed off to the left. The whole crew looked at the part of the sky where the man was excitedly pointing, and there was a UFO. "It didn't appear to be large," one of the scientists later said, "but it was plainly visible. It was easy to see that it was elliptical in shape and had a 'whitish-silver color.'" After taking a split second to realize what they were looking at, one of the men swung the theodolite around to pick up the object, and the timer reset his stop watch. For sixty seconds they tracked the UFO as it moved toward the east. In about fifty-five seconds it had dropped from an angle of elevation of 45 degrees to 25 degrees, then it zoomed upward and in a few seconds it was out of sight. The crew heard no sound and the New Mexico desert was so calm that day that they could have heard "a whisper a mile away."

When they reduced the data they had collected, McLaughlin and crew found out that the UFO had been traveling 4 degrees per second. At one time during the observed portion of its flight, the UFO had passed in front of a range of mountains that were visible to the observers. Using this as a check point, they estimated the size of the UFO to be 40 feet wide and 100 feet long, and they computed that the UFO had been at an altitude of 296,000 feet, or 56 miles, when they had first seen it, and that it was traveling 7 miles per second.

This wasn't the only UFO sighting made by White Sands scientists. On April 5, 1948, another team watched a UFO for several minutes as it streaked across the afternoon sky in a series of violent maneuvers. The disk-shaped object was about a fifth the size of a full moon.

On another occasion the crew of a C-47 that was tracking a skyhook balloon saw two similar UFO's come loping in from just above the horizon, circle the balloon, which was flying at just under 90,000 feet, and rapidly leave. When the balloon was recovered it was ripped.

I knew the two pilots of the C-47; both of them now believe in flying saucers. And they aren't alone; so do the people of the Aeronautical Division of General Mills who launch and track the big skyhook balloons. These scientists and engineers all have seen UFO's and they aren't their own balloons. I was almost tossed out of the General Mills offices into a cold January Minneapolis snowstorm for suggesting such a thing—but that comes later in our history of the UFO.

I don't know what these people saw. There has been a lot of interest generated by these sightings because of the extremely high qualifications and caliber of the observers. There is some legitimate doubt as to the accuracy of the speed and altitude figures that McLaughlin's crew arrived at from the data they measured with their theodolite. This doesn't mean much, however. Even if they were off by a factor of 100 per cent, the speeds and altitudes would be fantastic, and besides they looked at the UFO through a 25-power telescope and swore that it was a flat, oval-shaped object. Balloons, birds, and airplanes aren't flat and oval-shaped.

Astrophysicist Dr. Donald Menzel, in a book entitled *Flying Saucers*, says they saw a refracted image of their own balloon caused by an atmospheric phenomenon. Maybe he is right, but the General Mills people don't believe it. And their disagreement is backed up by years of practical experience with the atmosphere, its tricks and its illusions.

When the March issue of *True* magazine carrying Commander McLaughlin's story about how the White Sands Scientists had tracked UFO's reached the public, it stirred up a hornets' nest. Donald Keyhoe's article in the January *True* had converted many people but there were still a few heathens. The fact that government scientists had seen UFO's, and were admitting it, took care of a large percentage of these heathens. More and more people were believing in flying saucers.

The Navy had no comment to make about the sightings, but they did comment on McLaughlin. It seems that several months before, at the suggestion of a group of scientists at White Sands, McLaughlin had carefully written up the details of the sightings and forwarded them to Washington. The report contained no personal opinions, just facts. The comments on McLaughlin's report had been wired back to White Sands from Washington and they were, "What are you drinking out there?" A very intelligent answer—and it came from an admiral in the Navy's guided missile program.

By the time his story was published, McLaughlin was no longer at White Sands; he was at sea on the destroyer *Bristol*. Maybe he answered the admiral's wire.

The Air Force had no comment to make on McLaughlin's story. People at ATIC just shrugged and smiled as they walked by the remains of Project Grudge, and continued to "process UFO reports through regular intelligence channels."

In early 1950 the UFO's moved down to Mexico. The newspapers were full of reports. Tourists were bringing back more saucer stories than hand-tooled, genuine leather purses. *Time* reported that pickpockets were doing a fabulous business working the sky-gazing crowds that gathered when a *plativolo* was seen. Mexico's Department of National Defense reported that there had been some good reports but that the stories of finding crashed saucers weren't true.

On March 8 one of the best UFO sightings of 1950 took place right over ATIC.

About midmorning on this date a TWA airliner was coming in to land at the Dayton Municipal Airport. As the pilot circled to get into the traffic pattern, he and his copilot saw a bright light hovering off to the southeast. The pilot called the tower operators at the airport to tell them about the light, but before he could say anything, the tower operators told him they were looking at it too. They had called the operations office of the Ohio Air National Guard, which was located at the airport, and while the tower operators were talking, an Air Guard pilot was running toward an F-51, dragging his parachute, helmet, and oxygen mask.

I knew the pilot, and he later told me, "I wanted to find out once and for all what these screwy flying saucer reports were all about."

While the F-51 was warming up, the tower operators called ATIC and told them about the UFO and where to look to see it. The people at ATIC rushed out and there it was—an extremely bright light, much brighter and larger than a star. Whatever it was, it was high because every once in a while it would be blanked out by the thick, high, scattered clouds that were in the area. While the group of people were standing in front of ATIC watching the light, somebody ran in and called the radar lab at Wright Field to see if they had any radar "on the air." The people in the lab said that they didn't have, but they could get operational in a hurry. They said they would search southeast of the field with their radar and suggested that ATIC send some people over. By the time the ATIC people arrived at the radar lab the radar was on the air and had a target in the same position as the light that everyone was looking at. The radar was also picking up the Air Guard F-51 and an F-51 that had been scrambled from Wright- Patterson. The pilots of the Air Guard '51 and the Wright-Patterson '51 could both see the UFO, and they were going after it. The master sergeant who was operating the radar called the F-51's on the radio, got them together and started to vector them toward the target. As the two airplanes climbed they kept up a continual conversation with the radar operator to make sure they were all after the same thing. For several minutes they could clearly see the UFO, but when they reached about 15,000 feet, the clouds moved in and they lost it. The pilots made a quick decision; since radar showed that they were getting closer to the target, they decided to spread out to keep from colliding with one another and to go up through the clouds. They went on instruments and in a few seconds they were in the cloud. It was much worse than they'd expected; the cloud was thick, and the airplanes were icing up fast. An F-51 is far from being a good instrument ship, but they stayed in their climb until radar called and said that they were close to the target; in fact, almost on it. The pilots had another hurried radio conference and decided that since the weather was so bad they'd better come down. If a UFO, or something, was in the clouds, they'd hit it before they could see it. So they made a wise decision; they dropped the noses of their airplanes and dove back down into the clear. They circled awhile but the clouds didn't break. In a few minutes the master sergeant on the radar reported that the target was fading fast. The F-51's went in and landed.

When the target faded on the radar, some of the people went outside to visually look for the UFO, but it was obscured by clouds, and the clouds stayed for an hour. When it finally did clear for a few minutes, the UFO was gone.

A conference was held at ATIC that afternoon. It included Roy James, ATIC's electronics specialist and expert on radar UFO's. Roy had been over at the radar lab and had seen the UFO on the scope but neither the F-51 pilots nor the master sergeant who operated the radar were at the conference. The records show that at this meeting a unanimous decision was reached as to the identity of the UFO's. The bright light was Venus since Venus was in the southeast during midmorning on March 8, 1950, and the radar return was caused by the ice-laden cloud that the F-51 pilots had encountered. Ice-laden clouds can cause a radar return. The group of intelligence specialists at the meeting decided that this was further proved by the fact that as the F-51's approached the center of the cloud their radar return appeared to approach the UFO target on the radarscope. They were near the UFO and near ice, so the UFO must have been ice.

The case was closed.

I had read the report of this sighting but I hadn't paid too much attention to it because it had been "solved." But one day almost two years later I got a telephone call at my office at Project Blue Book. It was a master sergeant, the master sergeant who had been operating the radar at the lab. He'd just heard that the Air Force was again seriously investigating UFO's and he wanted to see what had been said about the Dayton Incident. He came over, read the report, and violently disagreed with what had been decided upon as the answer. He said that he'd been working with radar before World War II; he'd helped with the operational tests on the first microwave warning radars developed early in the war by a group headed by Dr. Luis Alvarez. He said that what he saw on that radarscope was no ice cloud; it was some type of aircraft. He'd seen every conceivable type of weather target on radar, he told me; thunderstorms, ice-laden clouds, targets caused by temperature inversions, and the works. They all had similar characteristics—the target was "fuzzy" and varied in intensity. But in this case the target was a good, solid return and he was convinced that it was caused by a good, solid object. And besides, he said, when the target began to fade on his scope he had raised the tilt of the antenna and the target came back, indicating that whatever it was, it was climbing. Ice-laden clouds don't climb, he commented rather bitterly.

Nor did the pilot of one of the F-51's agree with the ATIC analysis. The pilot who had been leading the two-ship flight of F-51's on that day told me that what he saw was no planet. While he and his wing man were climbing, and before the clouds obscured it, they both got a good look at the UFO, and it was getting bigger and more distinct all the time. As they climbed, the light began to take on a shape; it was definitely round. And if it had been Venus it should have been in the same part of the sky the next day, but the pilot said that he'd looked and it wasn't there. The ATIC report doesn't mention this point.

I remember asking him a second time what the UFO looked like; he said, "huge and metallic"—shades of the Mantell Incident.

The Dayton Incident didn't get much of a play from the press because officially it wasn't an unknown and there's nothing intriguing about an ice cloud and Venus. There were UFO reports in the newspapers, however.

One story that was widely printed was about a sighting at the naval air station at Dallas, Texas. Just before noon on March 16, Chief Petty Officer Charles Lewis saw a disk-shaped UFO come streaking across the sky and buzz a high-flying B-36. Lewis first saw the UFO coming in from the north, lower than the B-36; then he saw it pull up to the big bomber as it got closer. It hovered under the B-36 for an instant, then it went speeding off and disappeared. When the press inquired about the incident, Captain M. A. Nation, commander of the air station, vouched for his chief and added that the base tower operators had seen and reported a UFO to him about ten days before.

This story didn't run long because the next day a bigger one broke when the sky over the little town of Farmington, New Mexico, about 170 miles northwest of Albuquerque, was literally invaded by UFO's. Every major newspaper carried the story. The UFO's had apparently been congregating over the four corners area for two days because several people had reported seeing UFO's on March 15 and 16. But the seventeenth was the big day, every saucer this side of Polaris must have made a successful rendezvous over Farmington, because on that day most of the town's 3,600 citizens saw the mass fly-by. The first reports were made at 10:15A.M.; then for an hour the air was full of flying saucers. Estimates of the number varied from a conservative 500 to "thousands." Most all the observers said the UFO's were saucer- shaped, traveled at almost unbelievable speeds, and didn't seem to have any set flight path. They would dart in and out and seemed to avoid collisions only by inches. There was no doubt that they weren't hallucinations because the mayor, the local newspaper staff, ex- pilots, the highway patrol, and every type of person who makes up a community of 3,600 saw them.

I've talked to several people who were in Farmington and saw this now famous UFO display of St. Patrick's Day, 1950. I've heard dozens of explanations—cotton blowing in the wind, bugs' wings reflecting sunlight, a hoax to put Farmington on the map, and real honest-to- goodness flying saucers. One explanation was never publicized, however, and if there is an explanation, it is the best. Under certain conditions of extreme cold, probably 50 to 60 degrees below zero, the plastic bag of a skyhook balloon will get very brittle, and will take on the characteristics of a huge light bulb. If a sudden gust of wind or some other disturbance hits the balloon, it will shatter into a thousand pieces. As these pieces of plastic float down and are carried along by the wind, they could look like thousands of flying saucers.

On St. Patrick's Day a skyhook balloon launched from Holloman AFB, adjacent to the White Sands Proving Ground, did burst near Farmington, and it was cold enough at 60,000 feet to make the balloon brittle. True, the people at Farmington never found any pieces of plastic, but the small pieces of plastic are literally as light as feathers and could have floated far beyond the city.

The next day, on March 18, the Air Force, prodded by the press, shrugged and said, "There's nothing to it," but they had no explanation.

True magazine came through for a third time when their April issue, which was published during the latter part of March 1950, carried a roundup of

UFO photos. They offered seven photos as proof that UFO's existed. It didn't take a photo-interpretation expert to tell that all seven could well be of doubtful lineage, nevertheless the collection of photos added fuel to the already smoldering fire. The U.S. public was hearing a lot about flying saucers and all of it was on the pro side. For somebody who didn't believe in the things, the public thought that the Air Force was being mighty quiet.

The subject took on added interest on the night of March 26, when a famous news commentator said the UFO's were from Russia.

The next night Henry J. Taylor, in a broadcast from Dallas, Texas, said that the UFO's were Uncle Sam's own. He couldn't tell all he knew, but a flying saucer had been found on the beach near Galveston, Texas. It had USAF markings.

Two nights later a Los Angeles television station cut into a regular program with a special news flash; later in the evening the announcer said they would show the first photos of the real thing, our military's flying saucer. The photos turned out to be of the Navy XF-5-U, a World War II experimental aircraft that never flew.

The public was now thoroughly confused.

By now the words "flying saucer" were being batted around by every newspaper reporter, radio and TV newscaster, comedian, and man on the street. Some of the comments weren't complimentary, but as Theorem I of the publicity racket goes, "It doesn't make any difference what's said as long as the name's spelled right."

Early in April the publication that is highly revered by so many, *U.S. News and World Report*, threw in their lot. The UFO's belonged to the Navy. Up popped the old non-flying XF-5-U again.

Events drifted back to normal when Edward R. Murrow made UFO's the subject of one of his TV documentaries. He took his viewers around the U.S., talked to Kenneth Arnold, of original UFO fame, by phone and got the story of Captain Mantell's death from a reporter "who was there." Sandwiched in between accounts of actual UFO sightings were the pro and con opinions of top Washington brass, scientists, and the man on the street.

Even the staid New York *Times*, which had until now stayed out of UFO controversy, broke down and ran an editorial entitled, "Those Flying Saucers—Are They or Aren't They?"

All of this activity did little to shock the military out of their dogma. They admitted that the UFO investigation really hadn't been discontinued. "Any substantial reports of any unusual aerial phenomena would be processed through normal intelligence channels," they told the press.

Ever since July 4, 1947, ten days after the first flying saucer report, airline pilots had been reporting that they had seen UFO's. But the reports weren't frequent—maybe one every few months. In the spring of 1950 this changed, however, and the airline pilots began to make more and more reports—good reports. The reports went to ATIC but they didn't receive much attention. In a few instances there was a semblance of an investigation but it was half-hearted. The reports reached the newspapers too, and here they received a great deal more attention. The reports were investigated, and the stories checked and rechecked. When airline crews began to turn in one UFO report after another, it was difficult to believe the old "hoax, hallucination, and misidentification of known objects" routine. In April, May, and June of 1950 there were over thirty-five good reports from airline crews.

One of these was a report from a Chicago and Southern crew who were flying a DC-3 from Memphis to Little Rock, Arkansas, on the night of March 31. It was an exceptionally clear night, no clouds or haze, a wonderful night to fly. At exactly nine twenty-nine by the cockpit clock the pilot, a Jack Adams, noticed a white light off to his left. The copilot, G. W. Anderson, was looking at the chart but out of the corner of his eye he saw the pilot lean forward and look out the window, so he looked out too. He saw the light just as the pilot said, "What's that?"

The copilot's answer was classic: "No, not one of those things."

Both pilots had only recently voiced their opinions regarding the flying saucers and they weren't complimentary.

As they watched the UFO, it passed across the nose of their DC-3 and they got a fairly good look at it. Neither the pilot nor the copilot was positive of the object's shape because it was "shadowy" but they assumed it was disk-shaped because of the circular arrangement of eight or ten "portholes," each one glowing from a strong bluish-white light that seemed to come from the inside of whatever it was that they saw. The UFO also had a blinking white light on top, a fact that led many people to speculate that this UFO was another airliner. But this idea was quashed when it was announced that there were no other airliners in the area. The crew of the DC-3, when questioned on this possibility, were definite in their answers. If it had been another airplane, they could have read the number, seen the passengers, and darn near reached out and slugged the pilot for getting so close to them.

About a month later, over northern Indiana, TWA treated all the passengers of one of their DC-3 nights to a view of a UFO that looked like a "big glob of molten metal."

The official answer for this incident is that the huge orange-red UFO was nothing more than the light from the many northern Indiana blast furnaces reflecting a haze layer. Could be, but the pilots say no.

There were similar sightings in North Korea two years later—and FEAF Bomber Command had caused a shortage of blast furnaces in North Korea.

UFO sightings by airline pilots always interested me as much as any type of sighting. Pilots in general should be competent observers simply because they spend a large part of their lives looking around the sky. And pilots do look; one of the first things an aviation cadet is taught is to "Keep your head on a swivel"; in other words, keep looking around the sky. Of all the pilots, the airline pilots are the cream of this group of good observers. Possibly some second lieutenant just out of flying school could be confused by some unusual formation of ground lights, a meteor, or a star, but airline pilots have flown thousands of hours or they wouldn't be sitting in the left seat of an airliner, and they should be familiar with a host of unusual sights.

One afternoon in February 1953 I had an opportunity to further my study of UFO sightings by airline pilots. I had been out at Air Defense Command Headquarters in Colorado Springs and was flying back East on a United Airlines DC-6. There weren't many passengers on the airplane that afternoon but, as usual, the captain came strolling back through the cabin to chat. When he got to me he sat down in the next seat. We talked a few minutes; then I asked him what he knew about flying saucers. He sort of laughed and said that a dozen people a week asked that question, but when I told him who I was and why I was interested, his attitude changed. He said that he'd never seen a UFO but he knew a lot of pilots on United who had. One man, he told me, had seen one several years ago. He'd reported it but he had been sloughed off like the rest. But he was so convinced that he'd seen something unusual that he'd gone out and bought a Leica camera with a 105-mm. telephoto lens, learned how to use it, and now he carried it religiously during his flights.

There was a lull in the conversation, then the captain said, "Do you really want to get an opinion about flying saucers?"

I said I did.

"O.K.," I remember his saying, "how much of a layover do you have in Chicago?"

I had about two hours.

"All right, as soon as we get to Chicago I'll meet you at Caffarello's, across the street from the terminal building. I'll see who else is in and I'll bring them along."

I thanked him and he went back up front.

I waited around the bar at Caffarello's for an hour. I'd just about decided that he wasn't going to make it and that I'd better get back to catch my flight to Dayton when he and three other pilots came in. We got a big booth in the coffee shop because he'd called three more off-duty pilots who lived in Chicago and they were coming over too. I don't remember any of the men's names because I didn't make any attempt to. This was just an informal bull session and not an official interrogation, but I really got the scoop on what airline pilots think about UFO's.

First of all they didn't pull any punches about what they thought about the Air Force and its investigation of UFO reports. One of the men got right down to the point: "If I saw a flying saucer flying wing-tip formation with me and could see little men waving—even if my whole load of passengers saw it—I wouldn't report it to the Air Force."

Another man cut in, "Remember the thing Jack Adams said he saw down by Memphis?"

I said I did.

"He reported that to the Air Force and some red-hot character met him in Memphis on his next trip. He talked to Adams a few minutes and then told him that he'd seen a meteor. Adams felt like a fool. Hell, I know Jack Adams well and he's the most conservative guy I know. If he said he saw something with glowing portholes, he saw something with glowing portholes—and it wasn't a meteor."

Even though I didn't remember the pilots' names I'll never forget their comments. They didn't like the way the Air Force had handled UFO reports and I was the Air Force's "Mr. Flying Saucer." As quickly as one of the pilots would set me up and bat me down, the next one grabbed me off the floor and took his turn. But I couldn't complain too much; I'd asked for it. I think that this group of seven pilots pretty much represented the feelings of a lot of the airline pilots. They weren't wide-eyed space fans, but they and their fellow pilots had seen something and whatever they'd seen weren't hallucinations, mass hysteria, balloons, or meteors.

Three of the men at the Caffarello conference had seen UFO's or, to use their terminology, they had seen something they couldn't identify as a known object. Two of these men had seen odd lights closely following their airplanes at night. Both had checked and double- checked with CAA, but no other air-craft was in the area. Both admitted, however, that they hadn't seen enough to class what they'd seen as good UFO sighting. But the third man had a lulu.

If I recall correctly, this pilot was flying for TWA. One day in March 1952 he, his copilot, and a third person who was either a pilot deadheading home or another crew member, I don't recall which, were flying a C-54 cargo air-plane from Chicago to Kansas City. At about 2:30P.M. the pilot was checking in with the CAA radio at Kirksville, Missouri, flying 500 feet on top of a solid overcast. While he was talking he glanced out at his No. 2 engine, which had been losing oil. Directly in line with it, and a few degrees above, he saw a sil-very, disk-shaped object. It was too far out to get a really good look at it, yet it was close enough to be able definitely to make out the shape.

The UFO held its relative position with the C-54 for five or six minutes; then the pilot decided to do a little on-the-spot investigating himself. He started a gradual turn toward the UFO and for about thirty seconds he was getting closer, but then the UFO began to make a left turn. It had apparently slowed down because they were still closing on it.

About this time the copilot decided that the UFO was a balloon; it just looked as if the UFO was turning. The pilot agreed halfway—and since the company wasn't paying them to intercept balloons, they got back on their course to Kansas City. They flew on for a few more minutes with "the darn thing" still off to their left. If it was a balloon, they should be leaving it be-hind, the pilot recalled thinking to himself; if they made a 45-degree right turn, the "balloon" shouldn't stay off the left wing; it should drop 'way be-hind. So they made a 45-degree right turn, and although the "balloon" dropped back a little bit, it didn't drop back far enough to be a balloon. It seemed to put on speed to try to make a turn outside of the C-54's turn. The pilot continued on around until he'd made a tight 360-degree turn, and the UFO had followed, staying outside. They could not judge its speed, not knowing how far away it was, but to follow even a C-54 around in a 360-degree turn and to stay outside all of the time takes a mighty speedy object.

This shot the balloon theory right in the head. After the 360-degree turn the UFO seemed to be gradually losing altitude because it was getting below the level of the wings. The pilot decided to get a better look. He asked for full power on all four engines, climbed several thousand feet, and again turned into the UFO. He put the C-54 in a long glide, headed directly toward it. As they closed in, the UFO seemed to lose altitude a little faster and "sank" into the top of the overcast. Just as the C-54 flashed across the spot where the UFO had disappeared, the crew saw it rise up out of the overcast off their right wing and begin to climb so fast that in several seconds it was out of sight.

Both the pilot and copilot wanted to stay around and look for it but No. 2 engine had started to act up soon after they had put on full power for the climb, and they decided that they'd better get into Kansas City.

I missed my Dayton flight but I heard a good UFO story.

What had the two pilots and their passenger seen? We kicked it around plenty that afternoon. It was no balloon. It wasn't another airplane because when the pilot called Kirksville Radio he'd asked if there were any airplanes in the area. It might possibly have been a reflection of some kind except that when it "sank" into the overcast the pilot said it looked like something sinking into an overcast—it just didn't disappear as a reflection would. Then there was the sudden reappearance off the right wing. These are the types of things you just can't explain.

What did the pilots think it was? Three were sold that the UFO's were interplanetary spacecraft, one man was convinced that they were some U.S. "secret weapon," and three of the men just shook their heads. So did I. We all agreed on one thing—this pilot had seen something and it was something highly unusual.

The meeting broke up about 9:00P.M. I'd gotten the personal and very candid opinion of seven airline captains, and the opinions of half a hundred more airline pilots had been quoted. I'd learned that the UFO's are discussed often. I'd learned that many airline pilots take UFO sightings very seriously. I learned that some believe they are interplanetary, some think they're a U.S. weapon, and many just don't know. But very few are laughing off the good sightings.

By May 1950 the flying saucer business had hit a new all-time peak.

The Air Force didn't take any side, they just shrugged. There was no attempt to investigate and explain the various sightings. Maybe this was because someone was afraid the answer would be "Unknown." Or maybe it was because a few key officers thought that the eagles or stars on their shoulders made them leaders of all men. If they didn't believe in flying saucers and said so, it would be like calming the stormy Sea of Galilee. "It's all a bunch of damned nonsense," an Air Force colonel who was controlling the UFO investigation said. "There's no such thing as a flying saucer." He went on to say that all people who saw flying saucers were jokers, crackpots, or publicity hounds. Then he gave the airline pilots who'd been reporting UFO's a reprieve. "They were just fatigued," he said. "What they thought were spaceships were windshield reflections."

This was the unbiased processing of UFO reports through normal intelligence channels.

But the U.S. public evidently had more faith in the "crackpot" scientists who were spending millions of the public's dollars at the White Sands Proving Grounds, in the "publicity-mad" military pilots, and the "tired, old" airline pilots, because in a nationwide poll it was found that only 6 per cent of the country's 150,697,361 people agreed with the colonel and said, "There aren't such things."

Ninety-four per cent had different ideas.

Chapter Seven - The Pentagon Rumbles

On June 25, 1950, the North Korean armies swept down across the 38th parallel and the Korean War was on—the UFO was no longer a news item. But the lady, or gentleman, who first said, "Out of sight is out of mind," had never reckoned with the UFO.

On September 8, 1950, the UFO's were back in the news. On that day it was revealed, via a book entitled *Behind the Flying Saucers*, that government scientists had recovered and analyzed three different models of flying saucers. And they were fantastic— just like the book. They were made of an unknown super-duper metal and they were manned by little blue uniformed men who ate concentrated food and drank heavy water. The author of the book, Frank Scully, had gotten the story directly from a millionaire oilman, Silas Newton. Newton had in turn heard the story from an employee of his, a mysterious "Dr. Gee," one of the government scientists who had helped analyze the crashed saucers.

The story made news, Newton and "Dr. Gee" made fame, and Scully made money.

A little over two years later Newton and the man who was reportedly the mysterious "Dr. Gee" again made the news. The Denver district attorney's office had looked into the pair's oil business and found that the pockets they were trying to tap didn't contain oil. According to the December 6, 1952, issue of the *Saturday Review*, the D.A. had charged the two men with a $50,000 con game. One of their $800,000 electronic devices for their oil explorations turned out to be a $4.00 piece of war surplus junk.

Another book came out in the fall of 1950 when Donald Keyhoe expanded his original UFO story that had first appeared in the January 1950 issue of

True magazine. Next to Scully's book Keyhoe's book was tame, but it convinced more people. Keyhoe had based his conjecture on fact, and his facts were correct, even if the conjecture wasn't.

Neither the seesaw advances and retreats of the United Nations troops in Korea nor the two flying saucer books seemed to have any effect on the number of UFO reports logged into ATIC, however. By official count, seventy-seven came in the first half of 1950 and seventy-five during the latter half. The actual count could have been more because in 1950, UFO reports were about as popular as sand in spinach, and I would guess that at least a few wound up in the "circular file."

In early January 1951 I was recalled to active duty and assigned to Air Technical Intelligence Center as an intelligence officer. I had been at ATIC only eight and a half hours when I first heard the words "flying saucer" officially used. I had never paid a great deal of attention to flying saucer reports but I had read a few—especially those that had been made by pilots. I'd managed to collect some 2,000 hours of flying time and had seen many odd things in the air, but I'd always been able to figure out what they were in a few seconds. I was convinced that if a pilot, or any crew member of an airplane, said that he'd seen something that he couldn't identify he meant it—it wasn't a hallucination. But I wasn't convinced that flying saucers were spaceships.

My interest in UFO's picked up in a hurry when I learned that ATIC was the government agency that was responsible for the UFO project. And I was really impressed when I found out that the person who sat three desks down and one over from mine was in charge of the whole UFO show. So when I came to work on my second morning at ATIC and heard the words "flying saucer report" being talked about and saw a group of people standing around the chief of the UFO project's desk I about sprung an eardrum listening to what they had to say. It seemed to be a big deal—except that most of them were laughing. It must be a report of hoax or hallucination, I remember thinking to myself, but I listened as one of the group told the others about the report.

The night before a Mid-Continent Airlines DC-3 was taxiing out to take off from the airport at Sioux City, Iowa, when the airport control tower operators noticed a bright bluish-white light in the west. The tower operators, thinking that it was another airplane, called the pilot of the DC-3 and told him to be careful since there was another airplane approaching the field. As the DC-3 lined up to take off, both the pilots of the airliner and the tower operators saw the light moving in, but since it was still some distance away the DC-3 was given permission to take off. As it rolled down the runway getting up speed, both the pilot and the copilot were busy, so they didn't see the light approaching. But the tower operators did, and as soon as the DC-3 was airborne, they called and told the pilot to be careful. The copilot said that he saw the light and was watching it. Just then the tower got a call from another airplane that was requesting landing instructions and the operators looked away from the light.

In the DC-3 the pilot and copilot had also looked away from the light for a few seconds. When they looked back, the bluish-white light had apparently closed in because it was much brighter and it was dead ahead. In a split second it closed in and flashed by their right wing—so close that both pilots thought that they would collide with it. When it passed the DC-3, the pilots saw more than a light— they saw a huge object that looked like the "fuselage of a B-29."

When the copilot had recovered he looked out his side window to see if he could see the UFO and there it was, flying formation with them. He yelled at the pilot, who leaned over and looked just in time to see the UFO disappear.

The second look confirmed the Mid-Continent crew's first impression— the object looked like a B-29 without wings. They saw nothing more, only a big "shadowy shape" and the bluish-white light—no windows, no exhaust.

The tower had missed the incident because they were landing the other airplane and the pilot and the copilot didn't have time to call them and tell them about what was going on. All the tower operators could say was that seconds after the UFO had disappeared the light that they had seen was gone.

When the airliner landed in Omaha, the crew filed a report that was forwarded to the Air Force. But this wasn't the only report that was filed; a full colonel from military intelligence had been a passenger on the DC-3. He'd seen the UFO too, and he was mighty impressed.

I thought that this was an interesting report and I wondered what the official reaction would be. The official reaction was a great big, deep belly laugh.

This puzzled me because I'd read that the Air Force was seriously investigating all UFO reports.

I continued to eavesdrop on the discussions about the report all day since the UFO expert was about to "investigate" the incident. He sent out a wire to Flight Service and found that there was a B-36 somewhere in the area of Sioux City at the time of the sighting, and from what I could gather he was trying to blame the sighting on the B- 36. When Washington called to get the results of the analysis of the sighting, they must have gotten the B-36 treatment because the case was closed.

I'd only been at ATIC two days and I certainly didn't class myself as an intelligence expert, but it didn't take an expert to see that a B-36, even one piloted by an experienced idiot, could not do what the UFO had done—buzz a DC-3 that was in an airport traffic pattern.

I didn't know it at the time but a similar event had occurred the year before. On the night of May 29, 1950, the crew of an American Airlines DC-6 had just taken off from Washington National Airport, and they were about seven miles west of Mount Vernon when the copilot suddenly looked out and yelled, "Watch it—watch it." The pilot and the engineer looked out to see a bluish-white light closing in on them from dead ahead. The pilot racked the DC-6 up in a tight right turn while the UFO passed by on the left "from eleven to seven o'clock" and a little higher than the airliner. During this time the UFO passed between the full moon and DC-6 and the crew could see the dark silhouette of a "wingless B-29." Its length was about half the diameter of the full moon, and it had a blue flame shooting out the tail end.

Seconds after the UFO had passed by the DC-6, the copilot looked out and there it was again, apparently flying formation off their right wing. Then in a flash of blue flame it was gone—streaking out ahead of the airliner and making a left turn toward the coast.

The pilot of the DC-6, who made the report, had better than 15,000 hours' flying time.

I didn't hear anything about UFO's, or flying saucers, as they were then known, for several weeks but I kept them in mind and one day I asked one of the old hands at ATIC about them—specifically I wanted to know about the Sioux City Incident. Why had it been sloughed off so lightly? His answer was typical of the official policy at that time. "One of these days all of these crazy pilots will kill themselves, the crazy people on the ground will be locked up, and there won't be any more flying saucer reports."

But after I knew the people at ATIC a little better, I found that being anti-saucer wasn't a unanimous feeling. Some of the intelligence officers took the UFO reports seriously. One man, who had been on Project Sign since it was organized back in 1947, was convinced that the UFO's were interplanetary spaceships. He had questioned the people in the control tower at Godman AFB when Captain Mantell was killed chasing the UFO, and he had spent hours talking to the crew of the DC-3 that was buzzed near Montgomery, Alabama, by a "cigar-shaped UFO that spouted blue flame." In essence, he knew UFO history from *A to Z* because he had "been there."

I think that it was this controversial thinking that first aroused my interest in the subject of UFO's and led me to try to sound out a few more people.

The one thing that stood out to me, being unindoctrinated in the ways of UFO lore, was the schizophrenic approach so many people at ATIC took. On the surface they sided with the belly-laughers on any saucer issue, but if you were alone with them and started to ridicule the subject, they defended it or at least took an active interest. I learned this one day after I'd been at ATIC about a month.

A belated UFO report had come in from Africa. One of my friends was reading it, so I asked him if I could take a look at it when he had finished. In a few minutes he handed it to me.

When I finished with the report I tossed it back on my friend's desk, with some comment about the whole world's being nuts. I got a reaction I didn't expect; he wasn't so sure the whole world was nuts— maybe the nuts were at ATIC. "What's the deal?" I asked him. "Have they really thoroughly checked out every report and found that there's nothing to any of them?"

He told me that he didn't think so, he'd been at ATIC a long time. He hadn't ever worked on the UFO project, but he had seen many of their reports and knew what they were doing. He just plain didn't buy a lot of their explanations. "And I'm not the only one who thinks this," he added.

"Then why all of the big show of power against the UFO reports?" I remember asking him.

"The powers-that-be are anti-flying saucer," he answered about half bitterly, "and to stay in favor it behooves one to follow suit."

As of February 1951 this was the UFO project.

The words "flying saucer" didn't come up again for a month or two. I'd forgotten all about the two words and was deeply engrossed in making an analysis of the performance of the Mig-15. The Mig had just begun to show up in Korea, and finding out more about it was a hot project.

Then the words "flying saucer" drifted across the room once more. But this time instead of belly laughter there was a note of hysteria.

It seems that a writer from *Life* magazine was doing some research on UFO's and rumor had it that *Life* was thinking about doing a feature article. The writer had gone to the Office of Public Information in the Pentagon and had inquired about the current status of Project Grudge. To accommodate the writer, the OPI had sent a wire out to ATIC: What is the status of Project Grudge?

Back went a snappy reply: Everything is under control; each new report is being thoroughly analyzed by our experts; our vast files of reports are in tiptop shape; and in general things are hunky-dunky. All UFO reports are hoaxes, hallucinations, and the misidentification of known objects.

Another wire from Washington: Fine, Mr. Bob Ginna of *Life* is leaving for Dayton. He wants to check some reports.

Bedlam in the raw.

Other magazines had printed UFO stories, and other reporters had visited ATIC, but they had always stayed in the offices of the top brass. For some reason the name *Life*, the prospects of a feature story, and the feeling that this Bob Ginna was going to ask questions caused sweat to flow at ATIC.

Ginna arrived and the ATIC UFO "expert" talked to him. Ginna later told me about the meeting. He had a long list of questions about reports that had been made over the past four years and every time he asked a question, the "expert" would go tearing out of the room to try to find the file that had the answer. I remember that day people spent a lot of time ripping open bundles of files and pawing through them like a bunch of gophers. Many times, "I'm sorry, that's classified," got ATIC out of a tight spot.

Ginna, I can assure you, was not at all impressed by the "efficiently operating UFO project." People weren't buying the hoax, hallucination, and misidentification stories quite as readily as the Air Force believed.

Where it started or who started it I don't know, but about two months after the visit from *Life's* representative the official interest in UFO's began to pick up. Lieutenant Jerry Cummings, who had recently been recalled to active duty, took over the project.

Lieutenant Cummings is the type of person who when given a job to do does it. In a few weeks the operation of the UFO project had improved considerably. But the project was still operating under political, economic, and manpower difficulties. Cummings' desk was right across from mine, so I began to get a UFO indoctrination via bull sessions. Whenever Jerry found a good report in the pile—and all he had to start with was a pile of papers and files—he'd toss it over for me to read.

Some of the reports were unimpressive, I remember. But a few were just the opposite. Two that I remember Jerry's showing me made me wonder how the UFO's could be sloughed off so lightly. The two reports involved movies taken by Air Force technicians at White Sands Proving Ground in New Mexico.

The guided missile test range at White Sands is fully instrumented to track high, fast-moving objects—the guided missiles. Located over an area of many square miles there are camera stations equipped with cinetheodolite cameras and linked together by a telephone system.

On April 27, 1950, a guided missile had been fired, and as it roared up into the stratosphere and fell back to earth, the camera crews had recorded its flight. All the crews had started to unload their cameras when one of them spotted an object streaking across the sky. By April 1950 every person at White Sands was UFO-conscious, so one member of the camera crew grabbed a telephone headset, alerted the other crews, and told them to get pictures. Unfortunately only one camera had film in it, the rest had already been unloaded, and before they could reload, the UFO was gone. The photos from the one station showed only a smudgy dark object. About all the film proved was that something was in the air and whatever it was, it was moving.

Alerted by this first chance to get a UFO to "run a measured course," the camera crews agreed to keep a sharper lookout. They also got the official O.K. to "shoot" a UFO if one appeared.

Almost exactly a month later another UFO did appear, or at least at the time the camera crews thought that it was *a* UFO. This time the crews were ready—when the call went out over the telephone net that a UFO had been spotted, all of the crews scanned the sky. Two of the crews saw it and shot several feet of film as the shiny, bright object streaked across the sky.

As soon as the missile tests were completed, the camera crews rushed their film to the processing lab and then took it to the Data Reduction Group. But once again the UFO had eluded man because there were apparently two or more UFO's in the sky and each camera station had photographed a separate one. The data were no good for triangulation.

The records at ATIC didn't contain the analysis of these films but they did mention the Data Reduction Group at White Sands. So when I later took over the UFO investigation I made several calls in an effort to run down the actual film and the analysis. The files at White Sands, like all files, evidently weren't very good, because the original reports were gone. I did contact a major who was very co- operative and offered to try to find the people who had worked on the analysis of the film. His report, after talking to two men who had done the analysis, was what I'd expected—nothing concrete except that the UFO's

82

were unknowns. He did say that by putting a correction factor in the data gathered by the two cameras they were able to arrive at a rough estimate of speed, altitude, and size. The UFO was "higher than 40,000 feet, traveling over 2,000 miles per hour, and it was over 300 feet in diameter." He cautioned me, however, that these figures were only estimates, based on the possibly erroneous correction factor; therefore they weren't proof of anything—except that something was in the air.

The people at White Sands continued to be on the alert for UFO's while the camera stations were in operation because they realized that if the flight path of a UFO could be accurately plotted and timed it could be positively identified. But no more UFO's showed up.

One day Lieutenant Cummings came over to my desk and dropped a stack of reports in front of me. "All radar reports," he said, "and I'm getting more and more of them every day."

Radar reports, I knew, had always been a controversial point in UFO history, and if more and more radar reports were coming in, there was no doubt that an already controversial issue was going to be compounded.

To understand why there is always some disagreement whenever a flying saucer is picked up on radar, it is necessary to know a little bit about how radar operates.

Basically radar is nothing but a piece of electronic equipment that "shouts" out a radio wave and "listens" for the echo. By "knowing" how fast the radio, or radar, wave travels and from which direction the echo is coming, the radar tells the direction and distance of the object that is causing the echo. Any "solid" object like an airplane, bird, ship, or even a moisture-laden cloud can cause a radar echo. When the echo comes back to the radar set, the radar operator doesn't have to listen for it and time it because this is all done for him by the radar set and he sees the "answer" on his radarscope—a kind of a round TV screen. What the radar operator sees is a bright dot, called a "blip" or a "return." The location of the return on the scope tells him the location of the object that was causing the echo. As the object moves through the sky, the radar operator sees a series of bright dots on his scope that make a track. On some radar sets the altitude of the target, the object causing the echo, can also be measured.

Under normal conditions the path that the radar waves take as they travel through the air is known. Normal conditions are when the temperature and relative humidity of the air decrease with an increase in altitude. But sometimes a condition will occur where at some level, instead of the temperature and/or relative humidity decreasing with altitude, it will begin to increase. This layer of warm, moist air is known as an inversion layer, and it can do all kinds of crazy things to a radar wave. It can cause part of the radar wave to travel in a big arc and actually pick up the ground many miles away. Or it can cause the wave to bend down just enough to pick up trucks, cars, houses, or anything that has a surface perpendicular to the ground level.

One would immediately think that since the ground or a house isn't moving, and a car or truck is moving only 40, 50, or 60 miles an hour, a radar operator should be able to pick these objects out from a fast-moving target. But it isn't as simple as that. The inversion layer shimmers and moves, and one second the radar may be picking up the ground or a truck in one spot and the next second it may be picking up something in a different spot. This causes a series of returns on the scope and can give the illusion of extremely fast or slow speeds.

These are but a few of the effects of an inversion layer on radar. Some of the effects are well known, but others aren't. The 3rd Weather Group at Air Defense Command Headquarters in Colorado Springs has done a lot of work on the effects of weather on radar, and they have developed mathematical formulas for telling how favorable weather conditions are for "anomalous propagation," the two-bit words for false radar targets caused by weather.

The first problem in analyzing reports of UFO's being picked up on radar is to determine if the weather conditions are right to give anomalous propagation. This can be determined by putting weather data into a formula. If they are, then it is necessary to determine whether the radar targets were real or caused by the weather. This is the difficult job. In most cases the only answer is the appearance of the target on the radar-scope. Many times a weather target will be a fuzzy and indistinct spot on the scope while a real target, an airplane for example, will be bright and sharp. This question of whether a target looked real is the cause of the majority of the arguments about radar-detected UFO's because it is up to the judgment of the radar operator as to what the target looked like. And whenever human judgment is involved in a decision, there is plenty of room for an argument.

All during the early summer of 1951 Lieutenant Cummings "fought the syndicate" trying to make the UFO respectable. All the time I was continuing to get my indoctrination. Then one day with the speed of a shotgun wedding, the long-overdue respectability arrived. The date was September 12, 1951, and the exact time was 3:04P.M.

On this date and time a teletype machine at Wright-Patterson AFB began to chatter out a message. Thirty-six inches of paper rolled out of the machine before the operator ripped off the copy, stamped it Operational Immediate, and gave it to a special messenger to deliver to ATIC. Lieutenant Cummings got the message. The report was from the Army Signal Corps radar center at Fort Monmouth, New Jersey, and it was red-hot.

The incident had started two days before, on September 10, at 11:10A.M., when a student operator was giving a demonstration to a group of visiting brass at the radar school. He demonstrated the set under manual operation for a while, picking up local air traffic, then he announced that he would demonstrate automatic tracking, in which the set is put on a target and follows it without help from the operator. The set could track objects flying at jet speeds.

The operator spotted an object about 12,000 yards southeast of the station, flying low toward the north. He tried to switch the set to automatic tracking. He failed, tried again, failed again. He turned to his audience of VIPs, embarrassed.

"It's going too fast for the set," he said. "That means it's going faster than a jet!"

A lot of very important eyebrows lifted. What flies faster than a jet?

The object was in range for three minutes and the operator kept trying, without success, to get into automatic track. The target finally went off the scope, leaving the red-faced operator talking to himself. The radar technicians at Fort Monmouth had checked the weather—there wasn't the slightest indication of an inversion layer.

Twenty-five minutes later the pilot of a T-33 jet trainer, carrying an Air Force major as passenger and flying 20,000 feet over Point Pleasant, New Jersey, spotted a dull silver, disklike object far below him. He described it as 30 to 50 feet in diameter and as descending toward Sandy Hook from an altitude of a mile or so. He banked the T-33 over and started down after it. As he shot down, he reported, the object stopped its descent, hovered, then sped south, made a 120-degree turn, and vanished out to sea.

The Fort Monmouth Incident then switched back to the radar group. At 3:15P.M. they got an excited, almost frantic call from headquarters to pick up a target high and to the north—which was where the first "faster-than-a-jet" object had vanished—and to pick it up in a hurry. They got a fix on it and reported that it was traveling slowly at 93,000 feet. They also could see it visually as a silver speck.

What flies 18 miles above the earth?

The next morning two radar sets picked up another target that couldn't be tracked automatically. It would climb, level off, climb again, go into a dive. When it climbed it went almost straight up.

The two-day sensation ended that afternoon when the radar tracked another unidentified slow-moving object and tracked it for several minutes.

A copy of the message had also gone to Washington. Before Jerry could digest the thirty-six inches of facts, ATIC's new chief, Colonel Frank Dunn, got a phone call. It came from the office of the Director of Intelligence of the Air Force, Major General (now Lieutenant General) C. P. Cabell. General Cabell wanted somebody from ATIC to get to New Jersey—fast—and find out what was going on. As soon as the reports had been thoroughly investigated, the general said that he wanted a complete personal report. Nothing expedites like a telephone call from a general officer, so in a matter of hours Lieutenant Cummings and Lieutenant Colonel N. R. Rosengarten were on an airliner, New Jersey-bound.

The two officers worked around the clock interrogating the radar operators, their instructors, and the technicians at Fort Monmouth. The pilot who had chased the UFO in the T-33 trainer and his passenger were flown to New York, and they talked to Cummings and Rosengarten. All other radar stations in the area were checked, but their radars hadn't picked up anything unusual.

At about 4:00A.M. the second morning after they had arrived, the investigation was completed, Cummings later told. He and Lieutenant Colonel Rosengarten couldn't get an airliner out of New York in time to get them to the Pentagon by 10:00A.M., the time that had been set up for their report, so they chartered an airplane and flew to the capital to brief the general.

General Cabell presided over the meeting, and it was attended by his entire staff plus Lieutenant Cummings, Lieutenant Colonel Rosengarten, and a special representative from Republic Aircraft Corporation. The man from Republic supposedly represented a group of top U.S. industrialists and scientists who thought that there should be a lot more sensible answers coming from the Air Force regarding the UFO's. The man was at the meeting at the personal request of a general officer.

Every word of the two-hour meeting was recorded on a wire recorder. The recording was so hot that it was later destroyed, but not before I had heard it several times. I can't tell everything that was said but, to be conservative, it didn't exactly follow the tone of the official Air Force releases—many of the people present at the meeting weren't as convinced that the "hoax, hallucination, and misidentification" answer was

The first thing the general wanted to know was, "Who in hell has been giving me these reports that every decent flying saucer sighting is being investigated?"

Then others picked up the questioning.

"What happened to those two reports that General ——— sent in from Saudi Arabia? He saw those two flying saucers himself."

"And who released this big report, anyway?" another person added, picking up a copy of the Grudge Report and slamming it back down on the table.

Lieutenant Cummings and Lieutenant Colonel Rosengarten came back to ATIC with orders to set up a new project and report back to General Cabell when it was ready to go. But Cummings didn't get a chance to do much work on the new revitalized Project Grudge—it was to keep the old name—because in a few days he was a civilian. He'd been released from active duty because he was needed back at Cal Tech, where he'd been working on an important government project before his recall to active duty.

The day after Cummings got his separation orders, Lieutenant Colonel Rosengarten called me into his office. The colonel was chief of the Aircraft and Missiles branch and one of his many responsibilities was Project Grudge. He said that he knew that I was busy as group leader of my regular group but, if he gave me enough people, could I take Project Grudge? All he wanted me to do was to get it straightened out and operating; then I could go back to trying to outguess the Russians. He threw in a few comments about the good job I'd done straightening out other fouled-up projects. Good old "Rosy." With my ego sufficiently inflated, I said yes.

On many later occasions, when I'd land at home in Dayton just long enough for a clean clothes resupply, or when the telephone would ring at 2:00A.M. to report a new "hot" sighting and wake up the baby, Mrs. Ruppelt and I have soundly cussed my ego.

I had had the project only a few days when a minor flurry of good UFO reports started. It wasn't supposed to happen because the day after I'd taken over Project Grudge I'd met the ex-UFO "expert" in the hall and he'd nearly doubled up with laughter as he said something about getting stuck with Project Grudge. He predicted that I wouldn't get a report until the newspapers began to play up flying saucers again. "It's all mass hysteria," he said.

The first hysterical report of the flurry came from the Air Defense Command. On September 23, 1951, at seven fifty-five in the morning, two F-86's on an early patrol were approaching Long Beach, California, coming in on the west leg of the Long Beach Radio range. All of a sudden the flight leader called his ground controller—high at twelve o'clock he and his wing man saw an object. It was in a gradual turn to its left, and it wasn't another airplane. The ground controller checked his radars but they had nothing, so the ground controller called the leader of the F-86's back and told him to go after the object and try to identify it. The two airplanes started to climb.

By this time the UFO had crossed over them but it was still in a turn and was coming back. Several times they tried to intercept, but they could never climb up to it. Once in a while, when they'd appear to be getting close, the UFO would lazily move out of range by climbing slightly. All the time it kept orbiting to the left in a big, wide circle. After about ten minutes the flight leader told the ground controller, who had been getting a running account of the unsuccessful intercept, that their fuel was low and that they'd have to break off soon. They'd gotten a fairly good look at the UFO, the flight leader told the ground controller, and it appeared to be a silver airplane with highly swept-back wings. The controller acknowledged the message and said that he was scrambling all his alert airplanes from George AFB. Could the two F-86's stay in the area a few more minutes? They stayed and in a few minutes four more F- 86's arrived. They saw the UFO immediately and took over.

The two F-86's with nearly dry tanks went back to George AFB.

For thirty more minutes the newly arrived F-86's worked in pairs trying to get up to the UFO's altitude, which they estimated to be 55,000 feet, but they couldn't make it. All the time the UFO kept slowly circling and speeding up only when the F-86's seemed to get too close. Then they began to run out of fuel and asked for permission to break off the intercept.

By this time one remaining F-86 had been alerted and was airborne toward Long Beach. He passed the four homeward-bound F-86's as he was going in, but by the time he arrived over Long Beach the UFO was gone.

All the pilots except one reported a "silver airplane with highly swept-back wings." One pilot said the UFO looked round and silver to him.

The report ended with a comment by the local intelligence officer. He'd called Edwards AFB, the big Air Force test base north of Los Angeles, but they had nothing in the air. The officer concluded that the UFO was no airplane. In 1951 nothing we had would fly higher than the F-86.

This was a good report and I decided to dig in. First I had some more questions I wanted to ask the pilots. I was just in the process of formulating this set of questions when three better reports came in. They automatically got a higher priority than the Long Beach Incident.

Chapter Eight - The Lubbock Lights, Unabridged

When four college professors, a geologist, a chemist, a physicist, and a petroleum engineer, report seeing the same UFO's on fourteen different occasions, the event can be classified as, at least, unusual. Add the facts that hundreds of other people saw these UFO's and that they were photographed, and the story gets even better. Add a few more facts—that these UFO's were picked up on radar and that a few people got a close look at one of them, and the story begins to convince even the most ardent skeptics.

This was the situation the day the reports of the Lubbock Lights arrived at ATIC. Actually the Lubbock Lights, as Project Blue Book calls them, involved many widespread reports. Some of these incidents are known to the public, but the ones that added the emphasis and intrigue to the case and caused hundreds of hours of time to be spent analyzing the reports have not been told before. We collected all of these reports under the one title because there appeared to be a tie- in between them.

The first word of the sightings reached ATIC late in September 1951, when the mail girl dropped letters into my "in" basket. One of the letters was from Albuquerque, New Mexico, one was from a small town in Washington State, where I knew an Air Defense Command radar station was located, and the other from Reese AFB at Lubbock, Texas.

I opened the Albuquerque letter first. It was a report from 34th Air Defense at Kirtland AFB. The report said that on the evening of August 25, 1951, an employee of the Atomic Energy Commission's supersecret Sandia Corporation and his wife had seen a UFO. About dusk they were sitting in the back yard of their home on the outskirts of Albuquerque. They were gazing at the night sky, commenting on how beautiful it was, when both of them were startled at the sight of a huge airplane flying swiftly and silently over their home. The airplane had been in sight only a few seconds but they had gotten a good look at it because it was so low. They estimated 800 to 1,000 feet. It was the shape of a "flying wing" and one and a half times the size of a B-36. The wing was sharply swept back, almost like a V. Both the husband and wife had seen B-36's over their home many times. They couldn't see the color of the UFO but they did notice that there were dark bands running across the wing from front to back. On the aft edge of the wings there were six to eight pairs of soft, glowing, bluish lights. The aircraft had passed over their house from north to south.

The report went on to say that an investigation had been made immediately. Since the object might have been a conventional airplane, air traffic was checked. A commercial airlines Constellation was 50 miles west of Albuquerque and an Air Force B-25 was south of the city, but there had been nothing over Albuquerque that evening. The man's background was checked. He had a "Q" security clearance. This summed up his character, oddballs don't get "Q" clearances. No one else had reported the UFO, but this could be explained by the fact the AEC employee and his wife lived in such a location that anything passing over their home from north to south wouldn't pass over or near very many other houses. A sketch of the UFO was enclosed in the report.

I picked up the letter from Lubbock next. It was a thick report, and from the photographs that were attached, it looked interesting. I thumbed through it and stopped at the photos. The first thing that struck me was the similarity between these photos and the report I'd just read. They showed a series of lights in a V shape, very similar to those described as being on the aft edge of the "flying wing" that was reported from Albuquerque. This was something unique, so I read the report in detail.

On the night of August 25, 1951, about 9:20P.M., just twenty minutes after the Albuquerque sighting, four college professors from Texas Technological College at Lubbock had observed a formation of soft, glowing, bluish-green lights pass over their home. Several hours later they saw a similar group of lights and in the next two weeks they saw at least ten more. On August 31 an amateur photographer had taken five photos of the lights. Also on the thirty-first two ladies had seen a large "aluminum-colored," "pear-shaped" object hovering near a road north of Lubbock. The report went into the details of these sightings and enclosed a set of the photos that had been taken.

This report, in itself, was a good UFO report, but the similarity to the Albuquerque sighting, both in the description of the object and the time that it was seen, was truly amazing.

I almost overlooked the report from the radar station because it was fairly short. It said that early on the morning of August 26, only a few hours after the Lubbock sighting, two different radars had shown a target traveling 900 miles per hour at 13,000 feet on a northwesterly heading. The target had been observed for six minutes and an F-86 jet interceptor had been scrambled but by the time the F- 86 had climbed into the air the target was gone. The last paragraph in the report was rather curt and to the point. It was apparently in anticipation of the comments the report would draw. It said that the target was not caused by weather. The officer in charge of the radar station and several members of his crew had been operating radar for seven years and they could recognize a weather target. This target was real.

I quickly took out a map of the United States and drew in a course line between Lubbock and the radar station. A UFO flying between these two points would be on a northwesterly heading and the times it was seen at the two places gave it a speed of roughly 900 miles per hour.

This was by far the best combination of UFO reports I'd ever read and I'd read every one in the Air Force's files.

The first thing I did after reading the reports was to rush a set of the Lubbock photos to the intelligence officer of the 34th Air Division in Albuquerque. I asked him to show the photos to the AEC employee and his wife without telling them what they were. I requested an answer by wire. Later the next day I received my answer: "Observers immediately said that this is what they saw on the night of 25 August. Details by airmail." The details were a sketch the man and his wife had made of a wing around the photo of the Lubbock Lights. The number of lights in the photo and the number of lights the two observers had seen on the wing didn't tally, but they explained this by saying that they could have been wrong in their estimate.

The next day I flew to Lubbock to see if I could find an answer to all of these mysterious happenings.

I arrived in Lubbock about 5:00P.M. and contacted the intelligence officer at Reese AFB. He knew that I was on my way and had already set up a meeting with the four professors. Right after dinner we met them.

If a group had been hand-picked to observe a UFO, we couldn't have picked a more technically qualified group of people. They were:

Dr. W. I. Robinson, Professor of Geology.

Dr. A. G. Oberg, Professor of Chemical Engineering.

Professor W. L. Ducker, Head of the Petroleum Engineering Department.

Dr. George, Professor of Physics.

This is their story:

On the evening of August 25 the four men were sitting in Dr. Robinson's back yard. They were discussing micrometeorites and drinking tea. They jokingly stressed this point. At nine-twenty a formation of lights streaked across the sky directly over their heads. It all happened so fast that none of them had a chance to get a good look. One of the men mentioned that he had always admonished his students for not being more observant; now he was in that spot. He and his colleagues realized they could remember only a few details of what they had seen. The lights were a weird bluish-green color and they were in a semicircular formation. They estimated that there were from fifteen to thirty separate lights and that they were moving from north to south. Their one wish at this time was that the lights would reappear. They did; about an hour later the lights went over again. This time the professors were a little better prepared. With the initial shock worn off, they had time to get a better look. The details they had remembered from the first flight checked. There was one difference; in this flight the lights were not in any orderly formation, they were just in a group.

The professors reasoned that if the UFO's appeared twice they might come back. Come back they did. The next night and apparently many times later, as the professors made twelve more observations during the next few weeks.

For these later sightings they added two more people to their observing team.

Being methodical, as college professors are, they made every attempt to get a good set of data. They measured the angle through which the objects traveled and timed them. The several flights they checked traveled through 90 degrees of sky in three seconds, or 30 degrees per second. The lights usually suddenly appeared 45 degrees above the northern horizon, and abruptly went out 45 degrees above the southern horizon. They always traveled in this north-to-south direction. Outside of the first flight, in which the objects were in a roughly semicircular formation, in none of the rest of the flights did they note any regular pattern. Two or three flights were often seen in one night.

They had tried to measure the altitude, with no success. First they tried to compare the lights to the height of clouds but the clouds were never near the lights, or vice versa. Next they tried a more elaborate scheme. They measured off a base line perpendicular to the objects' usual flight path. Friends of the professors made up two teams. Each of the two teams was equipped with elevation-measuring devices, and one team was stationed at each end of the base line. The two teams were linked together by two-way radios. If they sighted the objects they would track and time them, thus getting the speed and altitude.

Unfortunately neither team ever saw the lights. But the lights never seemed to want to run the course. The wives of some of the watchers claimed to have seen them from their homes in the city. This later proved to be a clue.

The professors were not the sole observers of the mysterious lights. For two weeks hundreds of other people for miles around Lubbock reported that they saw the same lights. The professors checked many of these reports against the times of the flights they had seen and recorded, and many checked out close. They attempted to question these observers as to the length of time they had seen the lights and angles at which they had seen them, but the professors learned what I already knew, people are poor observers.

Naturally there has been much discussion among the professors and their friends as to the nature of the lights. A few simple mathematical calculations showed that if the lights were very high they would be traveling very fast. The possibility that they were some natural phenomena was, of course, discussed and seriously considered. The professors did a lot of thinking and research and decided that if they were natural phenomena they were something altogether new. Dr. George, who has since died, studied the phenomena of the night sky during his years as a professor at the University of Alaska, and he had never seen or heard of anything like this before.

This was the professors' story. It was early in the morning when we returned to Reese AFB. I sat up a few more hours unsuccessfully trying to figure out what they had seen.

The next day I again met the intelligence officer and we went to talk to Carl Hart, Jr., the amateur photographer who had taken the pictures of the lights. Hart was a freshman at Texas Tech. His story was that on the night of August 31 he was lying in his bed in an upstairs room of the Hart home. He, like everyone else in Lubbock, had heard about the lights but he had never seen them. It was a warm night and his bed was pushed over next to an open window. He was looking out at the clear night sky, and had been in bed about a half hour, when he saw a formation of the lights appear in the north, cross an open patch of sky, and disappear over his house. Knowing that the lights might reappear as they had done in the past, he grabbed his loaded Kodak 35, set the lens and shutter at f 3.5 and one tenth of a second, and went out into the middle of the back yard. Before long his vigil was rewarded when the lights made a second pass. He got two pictures. A third formation went over a few minutes later and he got three more pictures. The next morning bright and early Hart said he took the roll of unexposed film to a friend who ran a photo-finishing shop. He explained that he did all of his film processing in this friend's lab. He told the friend about the pictures and they quickly developed them.

I stopped Hart at this point and asked why he didn't get more excited about what could be the biggest news photos of the century. He said that the lights had appeared to be so dim that he was sure he didn't have anything on the negatives; had he thought that he did have some good pictures he would have awakened his friend to develop the negatives right away.

When he developed the negatives and saw that they showed an image, his friend suggested that he call the newspaper. At first the paper wasn't interested but then they decided to run the photos. I later found out that they had done some checking of their own.

We went with Hart into his back yard to re-enact what had taken place. He described the lights as being the same dull, glowing bluish- green color as those seen by the professors. The formation was different, however. The lights Hart saw were always flying in a perfect V. He traced the path from where they appeared over some trees in the north, through an open patch of sky over the back yard, to a point where they disappeared over the house. From the flight path he pointed out, the lights had crossed about 120 degrees of open sky in four seconds. This 30-degree-per-second angular velocity corresponded to the professors' measured angular velocity.

We made arrangements to borrow Hart's negatives, thanked him for his information, and left.

Armed with a list of names of other observers of the mysterious lights, the intelligence officer and I started out to try to get a cross-section account of the other UFO sightings in the Lubbock area. All the stories about the UFO's were the same; various types of formations of dull bluish-green lights, generally moving north to south. A few people had variations. One lady saw a flying Venetian blind and another a flying double boiler. One point of interest was that very few claimed to have seen the lights before reading the professors' story in the paper, but this could get back to the old question, "Do people look up if they have no reason to do so?"

We talked to observers in nearby towns. Their stories were the same. Two of them, tower operators at an airport, reported that they had seen the lights on several occasions.

It was in one of these outlying towns, Lamesa, that we talked to an old gentleman, about eighty years old, who gave us a good lead. He had seen the lights and he had identified them. Ever since he had read the story in the papers he had been looking. One evening he and his wife were in their yard looking for the lights. All of a sudden two or three appeared. They were in view for several seconds, then they were gone. In a few minutes the lights did a repeat performance. The man admitted he had been scared. He broke off his story of the lights and launched into his background as a native Texan, with range wars, Indians, and stagecoaches under his belt. What he was trying to point out was that despite the range wars, Indians, and stagecoaches, he had been scared. His wife had been scared too. We had some difficulty getting back to the lights but we finally made it. The third time they came around, he said, one of the lights emitted a sound. It said, "Plover." The old gentleman had immediately identified it as a plover, a water bird about the size of a quail. Later that night, and on several other occasions, they had seen the same thing. After a few more hair-raising but interesting stories of the old west Texas, we left.

Our next stop was the federal game warden's office in Lubbock. We got the low-down on plovers. We explained our interest and the warden was very helpful. He had been around west Texas all of his life so he was familiar with wildlife. The oily white breast of a plover could easily reflect light, but plovers usually didn't travel in more than pairs, or three at the most. He had never seen or heard of them traveling in a flock of fifteen to thirty but, of course, this wasn't impossible. Ducks, yes, but probably not plovers. He did say that for some unknown reason there were more than the usual number of plovers in the area that fall.

I was anxious to get the negatives that Hart had lent us back to the photo lab at Wright Field, but I had one more call to make. I wanted to talk to the two ladies who had seen a strange object hovering near their car, but I also wanted to write my report before I left Lubbock. Two Air Force special investigators from Reese AFB offered to talk to the ladies, so I stayed at the air base and finished my report.

That night when the investigators came back, I got the story. They had spent the whole day talking to the ladies and doing a little discreet checking into their backgrounds.

The two ladies, a mother and her daughter, had left their home in Matador, Texas, 70 miles northeast of Lubbock, about twelve-thirty P.M. on August 31. They were driving along in their car when they suddenly noticed "a pear-shaped" object about 150 yards ahead of them. It was just off the side of the road, about 120 feet in the air. It was drifting slowly to the east, "less than the speed required to take off in a Cub airplane." They drove on down the road about 50 more yards, stopped, and got out of the car. The object, which they estimated to be the size of a B-29 fuselage, was still drifting along slowly. There was no sign of any exhaust blast and they heard no noise, but they did see a "porthole" in the side of the object. In a few seconds the object began to pick up speed and rapidly climb out of sight. As it climbed it seemed to have a tight spiraling motion.

The investigation showed that the two ladies were "solid citizens," with absolutely no talents, or reasons, for fabricating such a story. The daughter was fairly familiar with aircraft. Her husband was an Air Force officer then in Korea, and she had been living near air bases for several years. The ladies had said that the object was "drifting" to the east, which possibly indicated that it was moving with the wind, but on further investigation it was found that it was moving *into* the wind.

The two investigators had worked all day and hadn't come up with the slightest indication of an answer.

This added the final section to my now voluminous report on the Lubbock affair.

The next morning as I rode to the airport to catch an airliner back to Dayton I tried to put the whole puzzle together. It was hard to believe that all Fd heard was real. Did a huge flying wing pass over Albuquerque and travel 250 miles to Lubbock in about fifteen minutes? This would be about 900 miles per hour. Did the radar station in Washington pick up the same thing? I'd checked the distances on the big wall map in flight operations just before leaving Reese AFB. It was 1,300 miles from Lubbock to the radar site. From talking to people, we decided that the lights were apparently still around Lubbock at 11:20P.M. and the radar picked them up just after midnight. They would have had to be traveling about 780 miles per hour. This was fairly close to the 900-mile-per-hour speed clocked by the two radars. The photos of the Lubbock Lights checked with the description of what the AEC employee and his wife had seen in Albuquerque. Nobody in Lubbock, however, had reported seeing a "flying wing" with lights. All of this was swimming around in my mind when I stepped out of the staff car at the Lubbock airport.

My plane had already landed so I checked in at the ticket counter, picked up a morning paper, and ran out and got into the airplane. I sat down next to a man wearing a Stetson hat and cowboy boots. I soon found out he was a retired rancher from Lubbock.

On the front page of the paper was an account of a large meteor that had flashed across New Mexico, west Texas, and Oklahoma the night before. According to the newspaper account, it was very spectacular and had startled a good many people in Lubbock. I was interested in the story because I had seen this meteor. It was a spectacular sight and I could easily understand how such things could be called UFO's. My seat partner must have noticed that I was reading the story of the meteor because he commented that a friend of his, the man who had brought him to the airport, had seen it. We talked about the meteor. This led to a discussion of other odd happenings and left a perfect opening for him to bring up the Lubbock Lights. He asked me if I'd heard about them. I said that I had heard a few vague stories. I hoped that this would stave off any detailed accounts of stories I had been saturated with during the past five days, but it didn't. I heard all the details all over again.

As he talked on, I settled back in my seat waiting for a certain thing to happen. Pretty soon it came. The rancher hesitated and the tone of his voice changed to a half-proud, half-apologetic tone. I'd heard this transition many times in the past few months; he was going to tell about the UFO that he had seen. He was going to tell how he had seen the bluish-green lights. I was wrong; what he said knocked me out of my boredom.

The same night that the college professors had seen their formation of lights his wife had seen something. Nobody in Lubbock knew about the story, not even their friends. He didn't want anyone to think he and his wife were "crazy." He was telling me only because I was a stranger. Just after dark his wife had gone outdoors to take some sheets off the clothesline. He was inside the house reading the paper. Suddenly his wife had rushed into the house, as he told the story, "as white as the sheets she was carrying." As close as he could remember, he said, this was about ten minutes before the professors made their first sighting. He stopped at this point to tell me about his wife, she wasn't prone to be "flighty" and she "never made up tales." This character qualification was also standard for UFO storytellers. The reason his wife was so upset was that she had seen a large object glide swiftly and silently over the house. She said it looked like "an airplane without a body." On the back edge of the wing were pairs of glowing bluish lights. The Albuquerque sighting! He said he didn't have any idea what his wife had seen but he thought that it was an interesting story.

It *was* an interesting story. It hit me right between the eyes. I knew the rancher and his wife couldn't have possibly heard the Albuquerque couple's story, only they and a few Air Force people knew about it. The chances of two identical stories being made up were infinitesimal, especially since neither of them fitted the standard Lubbock Light description. I wondered how many other people in Lubbock, Albuquerque, or anywhere in the Southwest had seen a similar UFO during this period and hesitated to mention it.

I tried to get a few more facts from the rancher but he'd told me all he knew. At Dallas I boarded an airliner to Dayton and he went on to Baton Rouge, never knowing what he'd added to the story of the Lubbock Lights.

On the way to Dayton I figured out a plan of attack on the thousands of words of notes I'd taken. The best thing to do, I decided, was to treat each sighting in the Lubbock Light series as a separate incident. All of them seemed to be dependent upon each other for importance. If the objects that were reported in several of the incidents could be identified, the rest would merely become average UFO reports. The photographs taken by Carl Hart, Jr., became number one on the agenda.

As soon as I reached Dayton I took Hart's negatives to the Photo Reconnaissance Laboratory at Wright Field. This laboratory, staffed by the Air Force's top photography experts, did all of our analysis of photographs. They went right to work on the negatives and soon had a report.

There had originally been five negatives, but when we asked to borrow them Hart could only produce four. The negatives were badly scratched and dirty because so many people had handled them, so it was difficult to tell the actual photographic images from the dust spots and scratches. The first thing that the lab did was to look at each spot on the negatives to see if it was an actual photographic image. They found that the photos showed an inverted V formation of lights. In each photo the individual image of a light was badly blurred due to motion of the camera, but by careful scrutiny of each blurred image they were able to determine that the original lights that Hart had photographed were circular, near pinpoint sources of light. Like a bright star, or a distant light bulb. Next they made enlargements from the negatives and carefully plotted the position of each light in the formation.

In each photograph the individual lights in the formation shifted position according to a definite pattern.

One additional factor that was brought out in the report was that although the photos were taken on a clear night no images of the stars could be found in the background. This proved one thing, the lights, which were overexposed in the photograph, were a great deal brighter than the stars, or the lights affected the film more than the light from the stars.

This was all that the photos showed. It was impossible to determine the size of each image of the group, speed, or altitude.

The next thing was to try to duplicate what Hart said he had done. I enlisted the aid of several friends and we tried to photograph a moving light. When we were talking to Hart in Lubbock, he had taken us to his back yard, where he had shot the pictures. He had traced the flight path of fights across the sky. We had him estimate the speed by following an imaginary flight of lights across the sky. It came out to about four seconds. We had a camera identical to the one that Hart had used and set up a light to move at the same speed as the UFO's had flown. We tried to take photographs. In four seconds we could get only two poor shots. These were badly blurred, much worse than Hart's, due to the one-tenth-of-a-second shutter speed. We repeated our experiment several times, each time with the same results. This made a lot of people doubt the authenticity of Hart's photos.

With the completed photo lab report in my hands, I was still without an answer. The report was interesting but didn't prove anything. All I could do was to get opinions from as qualified sources as I could find. A physiologist at the Aeromedical Laboratory knocked out the timing theory immediately by saying that if Hart had been excited he could have easily taken three photos in four seconds if we could get two in four seconds in our experiment. Several professional photographers, one of them a top *Life* photographer, said that if Hart was familiar with his camera and was familiar with panning action shots, his photos would have shown much less blur than ours. I recalled what I heard about Hart's having photographed sporting events for the Lubbock newspaper. This would have called for a good panning technique.

The photographs didn't tally with the description of the lights that the professors had seen; in fact, they were firmly convinced that they were of "home manufacture." The professors had reported soft, glowing lights yet the photos showed what should have been extremely bright lights. Hart reported a perfect formation while the professors, except for the first flight, reported an unorderly group. There was no way to explain this disagreement in the arrangement of the lights. Of course, it wasn't impossible that on the night that Hart saw the lights they were flying in a V formation. The first time the professors saw them they were flying in a semicircle.

The intensity of the lights was difficult to explain. Again I went to the people in the Photo Reconnaissance Laboratory. I asked them if there was any possible situation that could cause this. They said yes. An intensely bright light source which had a color far over in the red end of the spectrum, bordering on infrared, could do it. The eye is not sensitive to such a light, it could appear dim to the eye yet be "bright" to the film. I asked them what kind of a light source would cause this. There were several things, if you want to speculate, they said, extremely high temperatures for one. But this was as far as they would go. We have nothing in this world that flies that appears dim to the eye yet will show bright on film, they said.

This ended the investigation of the photographs, and the investigation ended at a blank wall. My official conclusion, which was later given to the press, was that "The photos were never proven to be a hoax but neither were they proven to be genuine." There is no definite answer.

The emphasis of the investigation was now switched to the professors' sighting. The meager amount of data that they had gathered seemed to be accurate but it was inconclusive as far as getting a definite answer was concerned. They had measured two things, how much of the sky the objects had crossed in a certain time and the angle from one side of the formation to the other. These figures didn't mean a great deal, however, since the altitude at which the formation of lights was flying was unknown. If you assumed that the objects were flying at an altitude of 10,000 feet you could easily compute that they were traveling about 3,600 miles per hour, or five to six times the speed of sound. The formation would have been about 1,750 feet wide. If each light was a separate object it could have been in the neighborhood of 100 feet in diameter. These figures were only a guess since nobody knew if the lights were at, above, or below 10,000 feet. If they had been higher they would have been going faster and have been larger. If lower than 10,000 feet, slower and smaller.

The only solid lead that had developed while the Reese AFB intelligence officer and I were investigating the professors' sightings was that the UFO's were birds reflecting the city lights; specifically plover. The old cowboy from Lamesa had described something identical to what the professors described

97

and they were plover. Secondly, whenever the professors left the vicinity of their homes to look for the lights they didn't see them, yet their wives, who stayed at home, did see them. If the "lights" were birds they would be flying low and couldn't be seen from more than a few hundred feet. While in Lubbock I'd noticed several main boulevards lighted with the bluish mercury vapor lights. I called the intelligence officer at Reese AFB and he airmailed me a city map of Lubbock with the mercury-vapor-lighted streets marked. The place where the professors had made their observations was close to one of these streets. The big hitch in this theory was that people living miles from a mercury-vapor-lighted boulevard had also reported the lights. How many of these sightings were due to the power of suggestion and how many were authentic I didn't know. If I could have found out, it would have been possible to plot the sightings in Lubbock, and if they were all located close to the lighted boulevards, birds would be an answer. This, however, it was impossible to do.

The fact that the lights didn't make any perceivable sound seemed as if it might be a clue. Birds or light phenomena wouldn't make any sound, but how about some object of appreciable size traveling at or above the speed of sound? Jet airplanes don't fly as fast as the speed of sound but they make a horrible roar. Artillery shells, which are going much faster than aircraft, whine as they go through the air. I knew that a great deal of the noise from a jet is due to the heated air rushing out of the tail pipe, but I didn't know exactly how much of the noise this caused. If a jet airplane with a silent engine could be built, how much noise would it make? How far could it be heard? To get the answer I contacted National Advisory Committee for Aeronautics Laboratory at Langley AFB, a government agency which specializes in aeronautical research. They didn't know. Neither they nor anybody else had ever done any research on this question. Their opinion was that such an aircraft could not be heard 5,000 or 10,000 feet away. Aerodynamicists at Wright Field's Aircraft Laboratory agreed.

I called the Army's Ballistic Research Laboratories at Aberdeen Proving Grounds, Maryland, to find out why artillery shells whine. These people develop and test all kinds of shells so they would have an answer if anybody did. They said that the majority of the whine of an artillery shell is probably caused by the flat back end of the shell. If a perfectly streamlined shell could be used it would not have any perceivable whine.

What I found out, or didn't find out, about the sound of an object moving at several times the speed of sound was typical of nearly every question that came up regarding UFO's. We were working in a field where there were no definite answers to questions. In some instances we were getting into fields far advanced above the then present levels of research. In other instances we were getting into fields where no research had been done at all. It made the problem of UFO analysis one of getting opinions. All we could do was hope the opinions we were getting were the best.

My attempts to reach a definite conclusion as to what the professors had seen met another blank wall. I had no more success than I'd had trying to reach a conclusion on the authenticity of the photographs.

A thorough analysis of the reports of the flying wings seen by the retired rancher's wife in Lubbock and the AEC employee and his wife in Albuquerque was made. The story from the two ladies who saw the aluminum-colored pear-shaped object hovering near the road near Matador, Texas, was studied, checked, and rechecked. Another blank wall on all three of these sightings.

By the time I got around to working on the report from the radar station in Washington State, the data of the weather conditions that existed on the night of the sighting had arrived. I turned the incident folder over to the electronics specialists at ATIC. They made the analysis and determined that the targets were caused by weather, although it was a borderline case. They further surmised that since the targets had been picked up on two radars, if I checked I'd find out that the two targets looked different on the two radarscopes. This is a characteristic of a weather target picked up on radars operating on different frequencies. I did check. I called the radar station and talked to the captain who was in charge of the crew the night the target had been picked up.

The target looked the same on both scopes. This was one of the reasons it had been reported, the captain told me. If the target hadn't been the same on both scopes, he wouldn't have made the report since he would have thought he had a weather target. He asked me what ATIC thought about the sighting. I said that Captain James thought it was weather. Just before the long-distance wires between Dayton and Washington melted, I caught some comment about people sitting in swivel chairs miles from the closest radarscope. . . . I took it that he didn't agree the target was caused by weather. But that's the way it officially stands today.

Although the case of the Lubbock Lights is officially dead, its memory lingers on. There have never been any more reliable reports of "flying wings" but lights somewhat similar to those seen by the professors have been reported. In about 70 per cent of these cases they were proved to be birds reflecting city lights.

The known elements of the case, the professors' sightings and the photos, have been dragged back and forth across every type of paper upon which written material appears, from the cheapest, coarsest pulp to the slick *Life* pages. Saucer addicts have studied and offered the case as all-conclusive proof, with photos, that UFO's are interplanetary. Dr. Donald Menzel of Harvard studied the case and ripped the sightings to shreds in *Look*, *Time*, and his book, *Flying Saucers*, with the theory that the professors were merely looking at refracted city lights. But none of these people even had access to the full report. This is the first time it has ever been printed.

The only other people outside Project Blue Book who have studied the complete case of the Lubbock Lights were a group who, due to their associations with the government, had complete access to our files. And these people were not pulp writers or wide-eyed fanatics, they were scientists—rocket experts, nuclear physicists, and intelligence experts. They had banded together to study our UFO reports because they were convinced that some of the UFO's that were being reported were interplanetary spaceships and the Lubbock series was one of these reports. The fact that the formations of lights were in different shapes didn't bother them; in fact, it convinced them all the more that their ideas of how a spaceship might operate were correct.

This group of scientists believed that the spaceships, or at least the part of the spaceship that came relatively close to the earth, would have to have a highly swept-back wing configuration. And they believed that for propulsion and control the craft had a series of small jet orifices all around its edge. Various combinations of these small jets would be turned on to get various flight attitudes. The lights that the various observers saw differed in arrangement because the craft was flying in different flight attitudes.

(Three years later the Canadian Government announced that this was exactly the way that they had planned to control the flying saucer that they were trying to build. They had to give up their plans for the development of the saucer-like craft, but now the project has been taken over by the U.S. Air Force.)

This is the complete story of the Lubbock Lights as it is carried in the Air Force files, one of the most interesting and most controversial collection of UFO sightings ever to be reported to Project Blue Book. Officially all of the sightings, except the UFO that was picked up on radar, are unknowns.

Personally I thought that the professors' lights might have been some kind of birds reflecting the light from mercury-vapor street lights, but I was wrong. They weren't birds, they weren't refracted light, but they weren't spaceships. The lights that the professors saw—the backbone of the Lubbock Light series—have been positively identified as a very commonplace and easily explainable natural phenomenon.

It is very unfortunate that I can't divulge exactly the way the answer was found because it is an interesting story of how a scientist set up complete instrumentation to track down the lights and how he spent several months testing theory after theory until he finally hit upon the answer. Telling the story would lead to his identity and, in exchange for his story, I promised the man complete anonymity. But he fully convinced me that he had the answer, and after having heard hundreds of explanations of UFO's, I don't convince easily.

With the most important phase of the Lubbock Lights "solved"—the sightings by the professors—the other phases become only good UFO reports.

Chapter Nine - The New Project Grudge

While I was in Lubbock, Lieutenant Henry Metscher, who was helping me on Project Grudge, had been sorting out the many bits and pieces of information that Lieutenant Jerry Cummings and Lieutenant Colonel Rosengarten had brought back from Fort Monmouth, New Jersey, and he had the answers.

The UFO that the student radar operator had assumed to be traveling at a terrific speed because he couldn't lock on to it turned out to be a 400-mile-an-hour conventional airplane. He'd just gotten fouled up on his procedures for putting the radar set on automatic tracking. The sighting by the two officers in the T-33 jet fell apart when Metscher showed how they'd seen a balloon.

The second radar sighting of the series also turned out to be a balloon. The frantic phone call from headquarters requesting a reading on the object's altitude was to settle a bet. Some officers in headquarters had seen the balloon launched and were betting on how high it was.

The second day's radar sightings were caused by another balloon and weather—both enhanced by the firm conviction that there were some mighty queer goings on over Jersey.

The success with the Fort Monmouth Incident had gone to our heads and we were convinced that with a little diligent digging we'd be knocking off saucers like an ace skeet-shooter. With all the confidence in the world, I attacked the Long Beach Incident, which I'd had to drop to go to Lubbock, Texas. But if saucers could laugh, they were probably zipping through the stratosphere chuckling to themselves, because there was no neat solution to this one.

In the original report of how the six F-86's chased the high-flying UFO over Long Beach, the intelligence officer who made the report had said that he'd checked all aircraft flights, therefore this wasn't the answer.

The UFO could have been a balloon, so I sent a wire to the Air Force weather detachment at the Long Beach Municipal Airport. I wanted the track of any balloon that was in the air at 7:55A.M. on September 23, 1951. While I was waiting for the answers to my two wires, Lieutenant Metscher and I began to sort out old UFO reports. It was a big job because back in 1949, when the old Project Grudge had been disbanded, the files had just been dumped into storage bins. Hank and I now had four filing case drawers full of a heterogeneous mass of UFO reports, letters, copies of letters, and memos.

But I didn't get to do much sorting because the mail girl brought in a copy of a wire that had just arrived. It was a report of a UFO sighting at Terre Haute, Indiana. I read it and told Metscher that I'd quickly whip out an answer and get back to helping him sort. But it didn't prove to be that easy.

The report from Terre Haute said that on October 9, a CAA employee at Hulman Municipal Airport had observed a silvery UFO. Three minutes later a pilot, flying east of Terre Haute, had seen a similar object. The report lacked many details but a few phone calls filled me in on the complete story.

At 1:43P.M. on the ninth a CAA employee at the airport was walking across the ramp in front of the administration building. He happened to glance up at the sky—why, he didn't know—and out of the corner of his eye he caught a flash of light on the southeastern horizon. He stopped and looked at the sky where the flash of light had been but he couldn't see anything. He was just about to walk on when he noticed what he described as "a pinpoint" of light in the same spot where he'd seen the flash. In a second or two the "pinpoint" grew larger and it was obvious to the CAA man that something was approaching the airport at a terrific speed. As he watched, the object grew larger and larger until it flashed directly overhead and disappeared to the northwest. The CAA man said it all happened so fast and he was so amazed that he hadn't called anybody to come out of the nearby hangar and watch the UFO. But when he'd calmed down he remembered a few facts. The UFO had been in sight for about fifteen seconds and during this time it had passed from horizon to horizon. It was shaped like a "flattened tennis ball," was a bright silver color, and when it was directly overhead it was "the size of a 50-cent piece held at arm's length."

But this wasn't all there was to the report. A matter of minutes after the sighting a pilot radioed Terre Haute that he had seen a UFO. He was flying from Greencastle, Indiana, to Paris, Illinois, when just east of Paris he'd looked back and to his left. There, level with his airplane and fairly close, was a large silvery object, "like a flattened orange," hanging motionless in the sky. He looked at it a few seconds, then hauled his plane around in a tight left bank. He headed directly toward the UFO, but it suddenly began to pick up speed and shot off toward the northeast. The time, by the clock on his instrument panel, was 1:45P.M.—just two minutes after the sighting at Terre Haute.

When I finished calling I got an aeronautical chart out of the file and plotted the points of the sighting. The CAA employee had seen the UFO disappear over the northwestern horizon. The pilot had been flying from Greencastle, Indiana, to Paris, Illinois, so he'd have been flying on a heading of just a little less than 270 degrees, or almost straight west. He was just east of Paris when he'd first seen the UFO, and since he said that he'd looked back and to his left, the spot where he saw the UFO would be right at a spot where the CAA man had seen his UFO disappear. Both observers had checked their watches with radio time just after the sightings, so there couldn't be more than a few seconds' discrepancy. All I could conclude was that both had seen the same UFO.

I checked the path of every balloon in the Midwest. I checked the weather—it was a clear, cloudless day; I had the two observers' backgrounds checked and I even checked for air traffic, although I knew the UFO wasn't an

102

airplane. I researched the University of Dayton library for everything on daylight meteors, but this was no good. From the description the CAA employee gave, what he'd seen had been a clear-cut, distinct, flattened sphere, with no smoke trail, no sparks and no tail. A daylight meteor, so low as to be described as "a 50-cent piece held at arm's length," would have had a smoke trail, sparks, and would have made a roar that would have jolted the Sphinx. This one was quiet. Besides, no daylight meteor stops long enough to let an airplane turn into it.

Conclusion: Unknown.

In a few days the data from the Long Beach Incident came in and I started to put it together. A weather balloon had been launched from the Long Beach Airport, and it was in the vicinity where the six F-86's had made their unsuccessful attempt to intercept a UFO. I plotted out the path of the balloon, the reported path of the UFO, and the flight paths of the F-86's. The paths of the balloon and the F-86's were accurate, I knew, because the balloon was being tracked by radio fixes and the F-86's had been tracked by radar. At only one point did the paths of the balloon, UFO, and F-86's coincide. When the first two F-86's made their initial visual contact with the UFO they were looking almost directly at the balloon. But from then on, even by altering the courses of the F-86's, I couldn't prove a thing.

In addition, the weather observers from Long Beach said that during the period that the intercept was taking place they had gone outside and looked at their balloon; it was an exceptionally clear day and they could see it at unusually high altitudes. They didn't see any F-86's around it. And one stronger point, the balloon had burst about ten minutes before the F-86's lost sight of the UFO.

Lieutenant Metscher took over and, riding on his Fort Monmouth victory, tried to show how the pilots had seen the balloon. He got the same thing I did—nothing.

On October 27, 1951, the new Project Grudge was officially established. I'd written the necessary letters and had received the necessary endorsements. I'd estimated, itemized, and justified direct costs and manpower. I'd conferred, inferred, and referred, and now I had the money to operate. The next step was to pile up all this paper work as an aerial barrier, let the saucers crash into it, and fall just outside the door.

I was given a very flexible operating policy for Project Grudge because no one knew the best way to track down UFO's. I had only one restriction and that was that I wouldn't have my people spending time doing a lot of wild speculating. Our job would be to analyze each and every UFO report and try to find what we believed to be an honest, unbiased answer. If we could not identify the reported object as being a balloon, meteor, planet, or one of half a hundred other common things that are sometimes called UFO's, we would mark the folder "Unknown" and file it in a special file. At some later date, when we built up enough of these "Unknown" reports, we'd study them.

As long as I was chief of the UFO project, this was our basic rule. If anyone became anti-flying saucer and was no longer capable of making an unbiased evaluation of a report, out he went. Conversely anyone who became a believer was through. We were too busy during the initial phases of the project to speculate as to whether the unknowns were spaceships, space monsters, Soviet weapons, or ethereal visions.

I had to let three people go for being too pro or too con.

By the latter part of November 1951 I knew most of what had taken place in prior UFO projects and what I expected to do. The people in Project Sign and the old Project Grudge had made many mistakes. I studied these mistakes and profited by them. I could see that my predecessors had had a rough job. Mine would be a little bit easier because of the pioneering they had done.

Lieutenant Metscher and I had sorted out all of the pre-1951 files, refiled them, studied them, and outlined the future course of the new Project Grudge.

When Lieut. Colonel Rosengarten and Lieutenant Cummings had been at the Pentagon briefing Major General Cabell on the Fort Monmouth incidents, the general had told them to report back when the new project was formed and ready to go. We were ready to go, but before taking my ideas to the Pentagon, I thought it might be wise to try them out on a few other people to get their reaction. Colonel Frank Dunn, then chief of ATIC, liked this idea. We had many well-known scientists and engineers who periodically visited ATIC as consultants, and Colonel Dunn suggested that these people's opinions and comments would be valuable. For the next two weeks every visitor to ATIC who had a reputation as a scientist, engineer, or scholar got a UFO briefing.

Unfortunately the names of these people cannot be revealed because I promised them complete anonymity. But the list reads like a page from *Great Men of Science.*

Altogether nine people visited the project during this trial period. Of the nine, two thought the Air Force was wasting its time, one could be called indifferent, and six were very enthusiastic over the project. This was a shock to me. I had expected reactions that ranged from an extremely cold absolute zero to a mild twenty below. Instead I found out that UFO's were being freely and seriously discussed in scientific circles. The majority of the visitors thought that the Air Force had goofed on previous projects and were very happy to find out that the project was being re-established. All of the visitors, even the two who thought we were wasting our time, had good suggestions on what to do. All of them offered their services at any future time when they might be needed. Several of these people became very good friends and valuable consultants later on.

About two weeks before Christmas, in 1951, Colonel Dunn and I went to the Pentagon to give my report. Major General John A. Samford had replaced Major General Cabell as Director of Intelligence, but General Samford must have been told about the UFO situation because he was familiar with the general aspects of the problem. He had appointed his Assistant for Production, Brigadier General W. M. Garland, to ride herd on the project for him.

Colonel Dunn briefly outlined to General Samford what we planned to do. He explained our basic policy, that of setting aside the unknowns and not speculating on them, and he told how the scientists visiting ATIC had liked the plans for the new Project Grudge.

There was some discussion about the Air Force's and ATIC's responsibility for the UFO reports. General Garland stated, and it was later confirmed in writing, that the Air Force was solely responsible for investigating and evaluating all UFO reports. Within the Air Force, ATIC was the responsible agency. This in turn meant that Project Grudge was responsible for all UFO reports made by any branch of the military service. I started my briefing by telling General Samford and his staff about the present UFO situation.

The UFO reports had never stopped coming in since they had first started in June 1947. There was some correlation between publicity and the number of sightings, but it was not an established fact that reports came in only when the press was playing up UFO's. Just within the past few months the number of good reports had increased sharply and there had been no publicity.

UFO's were seen more frequently around areas vital to the defense of the United States. The Los Alamos-Albuquerque area, Oak Ridge, and White Sands Proving Ground rated high. Port areas, Strategic Air Command bases, and industrial areas ranked next. UFO's had been reported from every state in the Union and from every foreign country. The U.S. did not have a monopoly.

The frequency of the UFO reports was interesting. Every July there was a sudden increase in the number of reports and July was always the peak month of the year. Just before Christmas there was usually a minor peak.

The Grudge Report had not been the solution to the UFO problem. It was true that a large percentage of the reports were due to the "mis- identification of known objects"; people were seeing balloons, airplanes, planets, but this was not the final answer. There were a few hoaxes, hallucinations, publicity-seekers, and fatigued pilots, but reports from these people constituted less than 1 per cent of the total. Left over was a residue of very good and very "unexplainable" UFO sightings that were classified as unknown.

The quality of the reports was getting better, I told the officers; they contained more details that could be used for analysis and the details were more precise and accurate. But still they left much to be desired.

Every one of the nine scientists and engineers who had reviewed the UFO material at ATIC had made one strong point: we should give top priority to getting reasonably accurate measurements of the speed, altitude, and size of reported UFO's. This would serve two purposes. First, it would make it easy to sort out reports of common things, such as balloons, airplanes, etc. Second, and more important, if we could get even one fairly accurate measurement that showed that some object was traveling through the atmosphere at high speed, and that it wasn't a meteor, the UFO riddle would be much easier to solve.

I had worked out a plan to get some measured data, and I presented it to the group for their comments.

I felt sure that before long the press would get wind of the Air Force's renewed effort to identify UFO's. When this happened, instead of being mysterious about the whole thing, we would freely admit the existence of the new project, explain the situation thoroughly and exactly as it was, and say that all UFO reports made to the Air Force would be given careful consideration. In this way we would encourage more people to report what they were seeing and we might get some good data.

To further explain my point, I drew a sketch on a blackboard. Suppose that a UFO is reported over a fair-sized city. Now we may get one or two reports, and these reports may be rather sketchy. This does us no good—all we can conclude is that somebody saw something that he couldn't identify. But suppose fifty people from all over the city report the UFO. Then it would be profitable for us to go out and talk to these people, find out the time they saw the UFO, and where they saw it (the direction and height above the horizon). Then we might be able to use these data, work out a triangulation problem, and get a fairly accurate measurement of speed, altitude, and size.

Radar, of course, will give an accurate measurement of speed and altitude, I pointed out, but radar is not infallible. There is always the problem of weather. To get accurate radar data on a UFO, it is always necessary to prove that it wasn't weather that was causing the target. Radar is valuable, and we wanted radar reports, I said, but they should be considered only as a parallel effort and shouldn't take the place of visual sightings.

In winding up my briefing, I again stressed the point that, as of the end of 1951—the date of this briefing—there was no positive proof that any craft foreign to our knowledge existed. All recommendations for the reorganization of Project Grudge were based solely upon the fact that there were many incredible reports of UFO's from many very reliable people. But they were still just flying saucer reports and couldn't be considered scientific proof.

Everyone present at the meeting agreed—each had read or had been briefed on these incredible reports. In fact, two of the people present had seen UFO's.

Before the meeting adjourned, Colonel Dunn had one last question. He knew the answer, but he wanted it confirmed. "Does the United States have a secret weapon that is being reported as a UFO?"

The answer was a flat "No."

In a few days I was notified that my plan had been given the green light. I already had the plan written up in the form of a staff study so I sent it through channels for formal approval.

It had been obvious right from the start of the reorganization of Project Grudge that there would be questions that no one on my staff was technically competent to answer. To have a fully staffed project, I'd need an astronomer, a physicist, a chemist, a mathematician, a psychologist, and probably a dozen other specialists. It was, of course, impossible to have all of these people on my staff, so I decided to do the next best thing. I would set up a contract with some research organization who already had such people on their staff; then I would call on them whenever their services were needed.

I soon found a place that was interested in such a contract, and the day after Christmas, Colonel S. H. Kirkland, of Colonel Dunn's staff, and I left Dayton for a two-day conference with these people to outline what we wanted. Their organization cannot be identified by name because they are doing other highly secret work for the government. I'll call them Project Bear.

Project Bear is a large, well-known research organization in the Midwest. The several hundred engineers and scientists who make up their staff run from experts on soils to nuclear physicists. They would make these people available to me to assist Project Grudge on any problem that might arise from a UFO report. They did not have a staff astronomer or psychologist, but they agreed to get them for us on a subcontract basis. Besides providing experts in every field of science, they would make two studies for us; a study of how much a person can be expected to see and remember from a UFO sighting, and a statistical study of UFO reports. The end product of the study of the powers of observation of a UFO observer would be an interrogation form.

Ever since the Air Force had been in the UFO business, attempts had been made to construct a form that a person who had seen a UFO could fill out. Many types had been tried but all of them had major disadvantages. Project Bear, working with the psychology department of a university, would study all of the previous questionnaires, along with actual UFO reports, and try to come up with as near a perfect interrogation form as possible. The idea was to make the form simple and yet extract as much and as accurate data as possible from the observer.

The second study that Project Bear would undertake would be a statistical study of all UFO reports. Since 1947 the Air Force had collected about 650 reports, but if our plan to encourage UFO reports worked out the way we expected this number could increase tenfold. To handle this volume of reports, Project Bear said that they would set up a complete UFO file on IBM punch cards. Then if we wanted any bit of information from the files, it would be a matter of punching a few buttons on an IBM card-sorting machine, and the files would be sorted electronically in a few seconds. Approximately a hundred items pertaining to a UFO report would be put on each card. These items included everything from the time the UFO was seen to its position in the sky and the observer's personality. The items punched on the cards would correspond to the items on the questionnaires that Project Bear was going to develop.

Besides giving us a rapid method of sorting data, this IBM file would give us a modus operandi file. Our MO file would be similar to the MO files used by police departments to file the methods of operations of a criminal. Thus when we received a report we could put the characteristics of the reported

UFO on an IBM punch card, put it into the IBM machine, and compare it with the characteristics of other sightings that had known solutions. The answer might be that out of the one hundred items on the card, ninety-five were identical to previous UFO reports that ducks were flying over a city at night reflecting the city's lights.

On the way home from the meeting Colonel Kirkland and I were both well satisfied with the assistance we believed Project Bear could give to Project Grudge.

In a few days I again left ATIC, this time for Air Defense Command Headquarters in Colorado Springs, Colorado. I wanted to find out how willing ADC was to help us and what they could do. When I arrived I got a thorough briefing on the operations of ADC and the promise that they would do anything they could to help solve the UFO riddle.

All of this co-operation was something that I hadn't expected. I'd been warned by the people who had worked on Project Sign and the old Project Grudge that everybody hated the word UFO—I'd have to fight for everything I asked for. But once again they were wrong. The scientists who visited ATIC, General Samford, Project Bear, and now Air Defense Command couldn't have been more co-operative. I was becoming aware that there was much wider concern about UFO reports than I'd ever realized before.

While I traveled around the United States getting the project set up, UFO reports continued to come in and all of them were good. One series of reports was especially good, and they came from a group of people who had had a great deal of experience watching things in the sky—the people who launch the big skyhook balloons for General Mills, Inc. The reports of what the General Mills people had seen while they were tracking their balloons covered a period of over a year. They had just sent them in because they had heard that Project Grudge was being reorganized and was taking a different view on UFO reports. They, like so many other reliable observers, had been disgusted with the previous Air Force attitude toward UFO reports, and they had refused to send in any reports. I decided that these people might be a good source of information, and I wanted to get further details on their reports, so I got orders to go to Minneapolis. A scientist from Project Bear went with me. We arrived on January 14, 1952, in the middle of a cold wave and a blizzard.

The Aeronautical Division of General Mills, Inc., of Wheaties and Betty Crocker fame, had launched and tracked every skyhook balloon that had been launched prior to mid-1952. They knew what their balloons looked like under all lighting conditions and they also knew meteorology, aerodynamics, astronomy, and they knew UFO's. I talked to these people for the better part of a full day, and every time I tried to infer that there might be some natural explanation for the UFO's I just about found myself in a fresh snowdrift.

What made these people so sure that UFO's existed? In the first place, they had seen many of them. One man told me that one tracking crew had seen so many that the sight of a UFO no longer even especially interested them. And the things that they saw couldn't be explained.

For example: On January 16, 1951, two people from General Mills and four people from Artesia, New Mexico, were watching a skyhook balloon from the Artesia airport. They had been watching the balloon off and on for about an hour when one of the group saw two tiny specks on the horizon, off to the northwest. He pointed them out to the others because two airplanes were expected into the airport, and he thought that these might be the airplanes. But as they watched, the two specks began to move in fast, and within a few seconds the observers could see that "the airplanes" were actually two round, dull white objects flying in close formation. The two objects continued to come in and headed straight toward the balloon. When they reached the balloon they circled it once and flew off to the northwest, where they disappeared over the horizon. As the two UFO's circled the balloon, they tipped on edge and the observers saw that they were disk-shaped.

When the two UFO's were near the balloon, the observers also had a chance to compare the size of the UFO's with the size of the balloon. If the UFO's were as close to the balloon as they appeared to be they would have been 60 feet in diameter.

After my visit to General Mills, Inc., I couldn't help remembering a magazine article I'd read about a year before. It said that there was not a single reliable UFO report that couldn't be attributed to a skyhook balloon.

I'd been back at ATIC only a few days when I found myself packing up to leave again. This time it was for New York. A high-priority wire had come into ATIC describing how a Navy pilot had chased a UFO over Mitchel AFB, on Long Island. It was a good report.

I remember the trip to New York because my train passed through Elizabeth, New Jersey, early in the morning, and I could see the fires caused by an American Airlines Convair that had crashed. This was the second of the three tragic Elizabeth, New Jersey, crashes.

The morning before, on January 21, a Navy pilot had taken off from Mitchel in a TBM. He was a lieutenant commander, had flown in World War II, and was now an engineer at the Navy Special Devices Center on Long Island. At nine-fifty he had cleared the traffic pattern and was at about 2,500 feet, circling around the airfield. He was southeast of the field when he first noticed an object below him and "about three runway lengths off the end of Runway 30." The object looked like the top of a parachute canopy, he told me; it was white and he thought he could see the wedges or panels. He said that he thought that it was moving across the ground a little bit too fast to be drifting with wind, but he was sure that somebody had bailed out and that he was looking at the top of his parachute. He was just ready to call the tower when he suddenly realized that this "parachute" was drifting across the wind. He had just taken off from Runway 30 and knew which direction the wind was blowing.

As he watched, the object, whatever it was (by now he no longer thought that it was a parachute), began to gradually climb, so he started to climb, he said, staying above and off to the right of the object. When the UFO started to make a left turn, he followed and tried to cut inside, but he overshot and passed over it. It continued to turn and gain speed, so he dropped the nose of the TBM, put on more power, and pulled in behind the object, which was now level with him. In a matter of seconds the UFO made a 180-degree turn and started to make a big swing around the northern edge of Mitchel AFB. The pilot tried to follow, but the UFO had begun to accelerate rapidly, and since a TBM leaves much to be desired on the speed end, he was getting farther and farther behind. But he did try to follow it as long as he could. As he made a wide turn around the northern edge of the airfield he saw that the UFO was now turning south. He racked the TBM up into a tight left turn to follow, but in a few seconds the UFO had disappeared. When he last saw it, it had crossed the Long Island coast line near Freeport and it was heading out to sea.

When he finished his account of the chase, I asked the commander some specific questions about the UFO. He said that just after he'd decided that the UFO was not a parachute it appeared to be at an altitude of about 200 to 300 feet over a residential section. From the time it took it to cover a city block, he'd estimated that it was traveling about 300 miles an hour. Even when he pulled in behind the object and got a good look, it still looked like a parachute canopy— dome-shaped—white—and it had a dark undersurface. It had been in sight two and a half minutes.

He had called the control tower at Mitchel during the chase, he told me, but only to ask if any balloons had been launched. He thought that he might be seeing a balloon. The tower had told him that there was a balloon in the area.

Then the commander took out an aeronautical chart and drew in his flight path and the apparent path of the UFO for me. I think that he drew it accurately because he had been continually watching landmarks as he'd chased the UFO and was very careful as he drew the sketches on the map.

I checked with the weather detachment at Mitchel and they said that they had released a balloon. They had released it at nine-fifty and from a point southeast of the airfield. I got a plot of its path. Just as in the Long Beach Incident, where the six F-86's tried to intercept the UFO, the balloon was almost exactly in line with the spot where the UFO was first seen, but then any proof you might attempt falls apart. If the pilot knew where he was, and had plotted his flight path even semi-accurately, he was never over the balloon. Yet he was over the UFO. He came within less than 2,000 feet of the UFO when he passed over it; yet he couldn't recognize it as a balloon even though he thought it might be a balloon since the tower had just told him that there was one in the area. He said that he followed the UFO around the north edge of the airfield. Yet the balloon, after it was launched southeast of the field, continued on a southeast course and never passed north of the airfield.

110

But the biggest argument against the object's being a balloon was the fact that the pilot pulled in behind it; it was directly off the nose of his airplane, and although he followed it for more than a minute, it pulled away from him. Once you line up an airplane on a balloon and go straight toward it you will catch it in a matter of seconds, even in the slowest airplane. There have been dogfights with UFO's where the UFO's turned out to be balloons, but the pilots always reported that the UFO "made a pass" at them. In other words, they rapidly caught up with the balloon and passed it. I questioned this pilot over and over on this one point, and he was positive that he had followed directly behind the UFO for over a minute and all the time it was pulling away from him.

This is one of the most typical UFO reports we had in our files. It is typical because no matter how you argue there isn't any definite answer. If you want to argue that the pilot didn't know where he was during the chase—that he was 3 or 4 miles from where he thought he was—that he never did fly around the northern edge of the field and get in behind the UFO—then the UFO could have been a balloon.

But if you want to believe that the pilot knew where he was all during the chase, and he did have several thousand hours of flying time, then all you can conclude is that the UFO was an unknown.

I think the pilot summed up the situation very aptly when he told me, "I don't know what it was, but I've never seen anything like it before or since—maybe it was a spaceship."

I went back to Dayton stumped—maybe it was a spaceship.

Chapter Ten - Project Blue Book and the Big Build-Up

Just twenty minutes after midnight on January 22, 1952, nineteen and a half hours after the Navy lieutenant commander had chased the UFO near Mitchel AFB, another incident involving an airplane and something unknown was developing in Alaska. In contrast with the unusually balmy weather in New York, the temperature in Alaska that night, according to the detailed account of the incident we received at ATIC, was a miserable 47 degrees below zero. The action was unfolding at one of our northernmost radar outposts in Alaska. This outpost was similar to those you may have seen in pictures, a collection of low, sprawling buildings grouped around the observatory- -like domes that house the antennae of the most modern radar in the world. The entire collection of buildings and domes are one color, solid white, from the plastering of ice and snow. The picture that the outpost makes could be described as fascinating, something out of a Walt Disney fantasy—but talk to somebody who's been there—it's miserable.

At 0020, twenty minutes after midnight, an airman watching one of the outpost's radarscopes saw a target appear. It looked like an airplane because it showed up as a bright, distinct spot. But it was unusual because it was northeast of the radar site, and very few airplanes ever flew over this area. Off to the northeast of the station there was nothing but ice, snow, and maybe a few Eskimos until you got to Russia. Occasionally a B-50 weather reconnaissance plane ventured into the area, but a quick check of the records showed that none was there on this night.

By the time the radar crew had gotten three good plots of the target, they all knew that it was something unusual—it was at 23,000 feet and traveling 1,500 miles an hour. The duty controller, an Air Force captain, was quickly called; he made a fast check of the targets that had now been put on the plotting board and called to a jet fighter-interceptor base for a scramble.

The fighter base, located about 100 miles south of the radar site, acknowledged the captain's call and in a matter of minutes an F-94 jet was climbing out toward the north.

While the F-94 was heading north, the radar crew at the outpost watched the unidentified target. The bright dots that marked its path had moved straight across the radarscope, passing within about 50 miles of the site. It was still traveling about 1,500 miles an hour. The radar had also picked up the F-94 and was directing it toward its target when suddenly the unidentified target slowed down, stopped, and reversed its course. Now it was heading directly toward the radar station. When it was within about 30 miles of the station, the radar operator switched his set to a shorter range and lost both the F-94 and the unidentified target.

While the radar operator was trying to pick up the target again, the F-94 arrived in the area. The ground controller told the pilot that they had lost the target and asked him to cruise around the area to see if he and his radar operator could pick up anything on the F-94's radar. The pilot said he would but that he was having a little difficulty, was low on fuel, and would have to get back to his base soon. The ground controller acknowledged the pilot's message, and called back to the air base telling them to scramble a second F-94.

The first F-94 continued to search the area while the ground radar tried to pick up the target but neither could find it.

About this time the second F-94 was coming in, so the ground radar switched back to long range. In a minute they had both of the F-94's and the unidentified target on their scope. The ground controller called the second F-94 and began to vector him into the target.

The first F-94 returned to its base.

As both the second F-94 and the target approached the radar site, the operator again switched to short range and again he lost the jet and the target. He switched back to long range, but by now they were too close to the radar site and he couldn't pick up either one.

The pilot continued on toward where the unidentified target should have been. Suddenly the F-94 radar operator reported a weak target off to the right at 28,000 feet. They climbed into it but it faded before they could make contact.

The pilot swung the F-94 around for another pass, and this time the radar operator reported a strong return. As they closed in, the F-94's radar showed that the target was now almost stationary, just barely moving. The F-94 continued on, but the target seemed to make a sudden dive and they lost it. The pilot of the jet interceptor continued to search the area but couldn't find anything. As the F-94 moved away from the radar station, it was again picked up on the ground radar, but the unidentified target was gone.

A third F-94 had been scrambled, and in the meantime its crew took over the search. They flew around for about ten minutes without detecting any targets on their radar. They were making one last pass almost directly over the radar station when the radar operator in the back seat of the F-94 yelled over the interphone that he had a target on his scope. The pilot called ground radar, but by this time both the F-94 and the unidentified target were again too close to the radar station and they couldn't be picked up. The F-94 closed in until it was within 200 yards of the target; then the pilot pulled up, afraid he might collide with whatever was out in the night sky ahead of him. He made another pass, and another, but each time the bright spot on the radar operator's scope just stayed in one spot as if something were defiantly sitting out in front of the F-94 daring the pilot to close in. The pilot didn't take the dare. On each pass he broke off at 200 yards.

The F-94 crew made a fourth pass and got a weak return, but it was soon lost as the target seemed to speed away. Ground radar also got a brief return, but in a matter of seconds they too lost the target as it streaked out of range on a westerly heading.

As usual, the first thing I did when I read this report was to check the weather. But there was no weather report for this area that was detailed enough to tell whether a weather inversion could have caused the radar targets.

But I took the report over to Captain Roy James, anyway, in hopes that he might be able to find a clue that would identify the UFO.

Captain James was the chief of the radar section at ATIC. He and his people analyzed all our reports where radar picked up UFO's. Roy had been familiar with radar for many years, having set up one of the first stations in Florida during World War II, and later he took the first aircraft control and warning squadron to Saipan. Besides worrying about keeping his radar operating, he had to worry about the Japs' shooting holes in his antennae.

Captain James decided that this Alaskan sighting I'd just shown him was caused by some kind of freak weather. He based his analysis on the fact that the unknown target had disappeared each time the ground radar had been switched to short range. This, he pointed out, is an indication that the radar was picking up some kind of a target that was caused by weather. The same weather that caused the ground radar to act up must have caused false targets on the F-94's radar too, he continued. After all, they had closed to within

200 yards of what they were supposedly picking up; it was a clear moonlight night, yet the crews of the F-94's hadn't seen a thing.

Taking a clue from the law profession, he quoted a precedent. About a year before over Oak Ridge, Tennessee, an F-82 interceptor had nearly flown into the ground three times as the pilot attempted to follow a target that his radar operator was picking up. There was a strong inversion that night, and although the target appeared as if it were flying in the air, it was actually a ground target.

Since Captain James was the chief of the radar section and he had said "Weather," weather was the official conclusion on the report. But reports of UFO's' being picked up on radar are controversial, and some of the people didn't agree with James's conclusion.

A month or two after we'd received the report, I was out in Colorado Springs at Air Defense Command Headquarters. I was eating lunch in the officers' club when I saw an officer from the radar operations section at ADC. He asked me to stop by his office when I had a spare minute, and I said that I would. He said that it was important.

It was the middle of the afternoon before I saw him and found out what he wanted. He had been in Alaska on TDY when the UFO had been picked up at the outpost radar site. In fact, he had made a trip to both the radar site and the interceptor base just two days after the sighting, and he had talked about the sighting with the people who had seen the UFO on the radar. He wanted to know what we thought about it.

When I told him that the sighting had been written off as weather, I remember that he got a funny look on his face and said, "Weather! What are you guys trying to pull, anyway?"

It was obvious that he didn't agree with our conclusion. I was interested in learning what this man thought because I knew that he was one of ADC's ace radar trouble shooters and that he traveled all over the world, on loan from ADC, to work out problems with radars.

"From the description of what the targets looked like on the radarscopes, good, strong, bright images, I can't believe that they were caused by weather," he told me.

Then he went on to back up his argument by pointing out that when the ground radar was switched to short range both the F-94 and the unknown target disappeared. If just the unknown target had disappeared, then it could have been weather. But since both disappeared, very probably the radar set wasn't working on short ranges for some reason. Next he pointed out that if there was a temperature inversion, which is highly unlikely in northern Alaska, the same inversion that would affect the ground radar wouldn't be present at 25,000 feet or above.

I told him about the report from Oak Ridge that Captain James had used as an example, but he didn't buy this comparison. At Oak Ridge, he pointed out, that F-82 was at only 4,000 feet. He didn't know how the F-94's could get to within 200 yards of an object without seeing it, unless the object was painted a dull black.

"No," he said, "I can't believe that those radar targets were caused by weather. I'd be much more inclined to believe that they were something real, something that we just don't know about."

During the early spring of 1952 reports of radar sightings increased rapidly. Most of them came from the Air Defense Command, but a few came from other agencies. One day, soon after the Alaskan Incident, I got a telephone call from the chief of one of the sections of a civilian experimental radar laboratory in New York State. The people in this lab were working on the development of the latest types of radar. Several times recently, while testing radars, they had detected unidentified targets. To quote my caller, "Some damn odd things are happening that are beginning to worry me." He went on to tell how the people in his lab had checked their radars, the weather, and everything else they could think of, but they could find absolutely nothing to account for the targets; they could only conclude that they were real. I promised him that his information would get to the right people if he'd put it in a letter and send it to ATIC. In about a week the letter arrived—hand-carried by no less than a general. The general, who was from Headquarters, Air Materiel Command, had been in New York at the radar laboratory, and he had heard about the UFO reports. He had personally checked into them because he knew that the people at the lab were some of the sharpest radar engineers in the world. When he found out that these people had already contacted us and had prepared a report for us, he offered to hand-carry it to Wright-Patterson.

I can't divulge how high these targets were flying or how fast they were going because it would give an indication of the performance of our latest radar, which is classified Secret. I can say, however, that they were flying mighty high and mighty fast.

I turned the letter over to ATIC's electronics branch, and they promised to take immediate action. They did, and really fouled it up. The person who received the report in the electronics branch was one of the old veterans of Projects Sign and Grudge. He knew all about UFO's. He got on the phone, called the radar lab, and told the chief (a man who possibly wrote all of the textbooks this person had used in college) all about how a weather inversion can cause false targets on weather. He was gracious enough to tell the chief of the radar lab to call if he had any more "trouble."

We never heard from them again. Maybe they found out what their targets were. Or maybe they joined ranks with the airline pilot who told me that if a flying saucer flew wing tip to wing tip formation with him, he'd never tell the Air Force.

In early February I made another trip to Air Defense Command Headquarters in Colorado Springs. This time it was to present a definite plan of how ADC could assist ATIC in getting better data on UFO's. I briefed General Benjamin W. Chidlaw, then the Commanding General of the Air Defense Command, and his staff, telling them about our plan. They agreed with it in principle and suggested that I work out the details with the Director of Intelligence for ADC, Brigadier General W. M. Burgess. General Burgess designated Major Verne Sadowski of his staff to be the ADC liaison officer with Project Grudge.

This briefing started a long period of close co-operation between Project Grudge and ADC, and it was a pleasure to work with these people. In all of my travels around the government, visiting and conferring with dozens of agencies, I never had the pleasure of working with or seeing a more smoothly operating and efficient organization than the Air Defense Command. General Chidlaw and General Burgess, along with the rest of the staff at ADC, were truly great officers. None of them were believers in flying saucers, but they recognized the fact that UFO reports were a problem that must be considered. With technological progress what it is today, you can't afford to have *anything* in the air that you can't identify, be it balloons, meteors, planets or flying saucers.

The plan that ADC agreed to was very simple. They agreed to issue a directive to all of their units explaining the UFO situation and telling specifically what to do in case one was detected. All radar units equipped with radarscope cameras would be required to take scope photos of targets that fell into the UFO category—targets that were not airplanes or known weather phenomena. These photos, along with a completed technical questionnaire that would be made up at ATIC by Captain Roy James, would be forwarded to Project Grudge.

The Air Defense Command UFO directive would also clarify the scrambling of fighters to intercept a UFO. Since it is the policy of the Air Defense Command to establish the identity of any unidentified target, there were no *special* orders issued for scrambling fighters to try to identify reported UFO's. A UFO was something unknown and automatically called for a scramble. However, there had been some hesitancy on the part of controllers to send airplanes up whenever radar picked up a target that obviously was not an airplane. The directive merely pointed out to the controllers that it was within the scope of existing regulations to scramble on radar targets that were plotted as traveling too fast or too slow to be conventional airplanes. The decision to scramble fighters was still up to the individual controller, however, and scrambling on UFO's would be a second or third priority.

The Air Defense Command UFO directive did not mention shooting at a UFO. This question came up during our planning meeting at Colorado Springs, but, like the authority to scramble, the authority to shoot at anything in the air had been established long ago. Every ADC pilot knows the rules for engagement, the rules that tell him when he can shoot the loaded guns that he always carries. If anything in the air over the United States commits any act that is covered by the rules for engagement, the pilot has the authority to open fire.

116

The third thing that ADC would do would be to integrate the Ground Observer Corps into the UFO reporting net. As a second priority, the GOC would report UFO's—first priority would still be reporting aircraft.

Ever since the new Project Grudge had been organized, we hadn't had to deal with any large-scale publicity about UFO's. Occasionally someone would bring in a local item from some newspaper about a UFO sighting, but the sightings never rated more than an inch or two column space. But on February 19, 1952, the calm was broken by the story of how a huge ball of fire paced two B-29's in Korea. The story didn't start a rash of reports as the story of the first UFO sighting did in June 1947, but it was significant in that it started a slow build-up of publicity that was far to surpass anything in the past.

This Korean sighting also added to the growing official interest in Washington. Almost every day I was getting one or two telephone calls from some branch of the government, and I was going to Washington at least once every two weeks. I was beginning to spend as much time telling people what was going on as I was doing anything about it. The answer was to get somebody in the Directorate of Intelligence in the Pentagon to act as a liaison officer. I could keep this person informed and he could handle the "branch office" in Washington. Colonel Dunn bought this idea, and Major Dewey J. Fournet got the additional duty of manager of the Pentagon branch. In the future all Pentagon inquiries went to Major Fournet, and if he couldn't answer them he would call me. The arrangement was excellent because Major Fournet took a very serious interest in UFO's and could always be counted on to do a good job.

Sometime in February 1952 I had a visit from two Royal Canadian Air Force officers. For some time, I learned, Canada had been getting her share of UFO reports. One of the latest ones, and the one that prompted the visit by the RCAF officers, occurred at North Bay, Ontario, about 250 miles north of Buffalo, New York. On two occasions an orange-red disk had been seen from a new jet fighter base in the area.

The Canadians wanted to know how we operated. I gave them the details of how we were currently operating and how we hoped to operate in the future, as soon as the procedures that were now in the planning stages could be put into operation. We agreed to try to set up channels so that we could exchange information and tie in the project they planned to establish with Project Grudge.

Our plans for continuing liaison didn't materialize, but through other RCAF intelligence officers I found out that their plans for an RCAF-sponsored project failed. A quasi-official UFO project was set up soon after this, however, and its objective was to use instruments to detect objects coming into the earth's atmosphere. In 1954 the project was closed down because during the two years of operation they hadn't officially detected any UFO's. My sources of information stressed the word "officially."

117

During the time that I was chief of the UFO project, the visitors who passed through my office closely resembled the international brigade. Most of the visits were unofficial in the sense that the officers came to ATIC on other business, but in many instances the other business was just an excuse to come out to Dayton to get filled in on the UFO story. Two RAF intelligence officers who were in the U.S. on a classified mission brought six single-spaced typed pages of questions they and their friends wanted answered. On many occasions Air Force intelligence officers who were stationed in England, France, and Germany, and who returned to the U.S. on business, took back stacks of unclassified flying saucer stories. One civilian intelligence agent who frequently traveled between the U.S. and Europe also acted as the unofficial courier for a German group— transporting hot newspaper and magazine articles about UFO's that I'd collected. In return I received the latest information on European sightings—sightings that never were released and that we never received at ATIC through official channels.

Ever since the fateful day when Lieutenant Jerry Cummings dropped his horn-rimmed glasses down on his nose, tipped his head forward, peered at Major General Cabell over his glasses and, acting not at all like a first lieutenant, said that the UFO investigation was all fouled up, Project Grudge had been gaining prestige. Lieutenant Colonel Rosengarten's promise that I'd be on the project for only a few months went the way of all military promises. By March 1952, Project Grudge was no longer just a project within a group; we had become a separate organization, with the formal title of the Aerial Phenomena Group. Soon after this step-up in the chain of command the project code name was changed to Blue Book. The word "Grudge" was no longer applicable. For those people who like to try to read a hidden meaning into a name, I'll say that the code name Blue Book was derived from the title given to college tests. Both the tests and the project had an abundance of equally confusing questions.

Project Blue Book had been made a separate group because of the steadily increasing number of reports we were receiving. The average had jumped from about ten a month to twenty a month since December 1951. In March of 1952 the reports slacked off a little, but April was a big month. In April we received ninety-nine reports.

On April 1, Colonel S. H. Kirkland and I went to Los Angeles on business. Before we left ATIC we had made arrangements to attend a meeting of the Civilian Saucer Investigators, a now defunct organization that was very active in 1952.

They turned out to be a well-meaning but Don Quixote-type group of individuals. As soon as they outlined their plans for attempting to solve the UFO riddle, it was obvious that they would fail. Project Blue Book had the entire Air Force, money, and enthusiasm behind it and we weren't getting any answers yet. All this group had was the enthusiasm.

The highlight of the evening wasn't the Civilian Saucer Investigators, however; it was getting a chance to read Ginna's UFO article in an advance copy of *Life* magazine that the organization had obtained—the article written from the material Bob Ginna had been researching for over a year. Colonel Kirkwood took one long look at the article, sidled up to me, and said, "We'd better get back to Dayton quick; you're going to be busy." The next morning at dawn I was sound asleep on a United Airlines DC-6, Dayton-bound.

The *Life* article undoubtedly threw a harder punch at the American public than any other UFO article ever written. The title alone, "Have We Visitors from Outer Space?" was enough. Other very reputable magazines, such as True, had said it before, but coming from *Life*, it was different. *Life* didn't say that the UFO's were from outer space; it just said maybe. But to back up this "maybe," it had quotes from some famous people. Dr. Walther Riedel, who played an important part in the development of the German V-2 missile and is presently the director of rocket engine research for North American Aviation Corporation, said he believed that the UFO's were from outer space. Dr. Maurice Biot, one of the world's leading aerodynamicists, backed him up.

But the most important thing about the *Life* article was the question in the minds of so many readers: "Why was it written?" *Life* doesn't go blasting off on flights of space fancy without a good reason. Some of the readers saw a clue in the author's comments that the hierarchy of the Air Force was now taking a serious look at UFO reports. "Did the Air Force prompt *Life* to write the article?" was the question that many people asked themselves.

When I arrived at Dayton, newspapermen were beating down the door. The official answer to the *Life* article was released through the Office of Public Information in the Pentagon: "The article is factual, but *Life's* conclusions are their own." In answer to any questions about the article's being Air Force-inspired, my weasel- worded answer was that we had furnished *Life* with some raw data on specific sightings.

My answer was purposely weasel-worded because I knew that the Air Force had unofficially inspired the *Life* article. The "maybe they're interplanetary" with the "maybe" bordering on "they are" was the personal opinion of several very high-ranking officers in the Pentagon—so high that their personal opinion was almost policy. I knew the men and I knew that one of them, a general, had passed his opinions on to Bob Ginna.

Oddly enough, the *Life* article did not cause a flood of reports. The day after the article appeared we got nine sightings, which was unusual, but the next day they dropped off again.

The number of reports did take a sharp rise a few days later, however. The cause was the distribution of an order that completed the transformation of the UFO from a bastard son to the family heir. The piece of paper that made Project Blue Book legitimate was Air Force Letter 200-5, Subject: Unidentified Flying Objects. The letter, which was duly signed and sealed by the Secretary of the Air Force, in essence stated that UFO's were not a joke, that the Air Force was making a serious study of the problem, and that Project Blue Book was responsible for the study. The letter stated that the commander of every Air Force installation was responsible for forwarding all UFO reports to ATIC by wire, with a copy to the Pentagon. Then a more detailed report would be sent by airmail. Most important of all, it gave Project Blue Book the authority to directly contact any Air Force unit in the United States without going through any chain of command. This was almost unheard of in the Air Force and gave our project a lot of prestige.

The new reporting procedures established by the Air Force letter greatly aided our investigation because it allowed us to start investigating the better reports before they cooled off. But it also had its disadvantages. It authorized the sender to use whatever priority he thought the message warranted. Some things are slow in the military, but a priority message is not one of them. When it comes into the message center, it is delivered to the addressee immediately, and for some reason, all messages reporting UFO's seemed to arrive between midnight and 4:00A.M. I was considered the addressee on all UFO reports. To complicate matters, the messages were usually classified and I would have to go out to the air base and personally sign for them.

One such message came in about 4:30A.M. on May 8, 1952. It was from a CAA radio station in Jacksonville, Florida, and had been forwarded over the Flight Service teletype net. I received the usual telephone call from the teletype room at Wright-Patterson, I think I got dressed, and I went out and picked up the message. As I signed for it I remember the night man in the teletype room said, "This is a lulu, Captain."

It was a lulu. About one o'clock that morning a Pan-American airlines DC-4 was flying south toward Puerto Rico. A few hours after it had left New York City it was out over the Atlantic Ocean, about 600 miles off Jacksonville, Florida, flying at 8,000 feet. It was a pitch-black night; a high overcast even cut out the glow from the stars. The pilot and copilot were awake but really weren't concentrating on looking for other aircraft because they had just passed into the San Juan Oceanic Control Area and they had been advised by radio that there were no other airplanes in the area. The copilot was turning around to look at number four engine when he noticed a light up ahead. It looked like the taillight of another airplane. He watched it closely for a few seconds since no other airplanes were supposed to be in the area. He glanced out at number four engine for a few seconds, looked back, and he saw that the light was in about the same position as when he'd first seen it. Then he looked down at the prop controls, synchronized the engines, and looked up again. In the few seconds that he had glanced away from the light, it had moved to the right so that it was now directly ahead of the DC-4, and it had increased in size. The copilot reached over and slapped the pilot on the shoulder and pointed. Just at that instant the light began to get bigger and bigger until it was "ten times the size of a landing light of an airplane." It continued to close in and with a flash it streaked by the DC-4's left wing. Before the crew could react and say anything, two more smaller balls of fire flashed by. Both pilots later said that they sat in their seats for several seconds with sweat trickling down their backs.

It was one of these two pilots who later said, "Were you ever traveling along the highway about 70 miles an hour at night, have the car that you were meeting suddenly swerve over into your lane and then cut back so that you just miss it by inches? You know the sort of sick, empty feeling you get when it's all over? That's just the way we felt."

As soon as the crew recovered from the shock, the pilot picked up his mike, called Jacksonville Radio, and told them about the incident. Minutes later we had the report. The next afternoon Lieutenant Kerry Rothstien, who had replaced Lieutenant Metscher on the project, was on his way to New York to meet the pilots when they returned from Puerto Rico.

When Kerry talked to the two pilots, they couldn't add a great deal to their original story. Their final comment was the one we all had heard so many times, "I always thought these people who reported flying saucers were crazy, but now I don't know."

When Lieutenant Rothstien returned to Dayton he triple-checked with the CAA for aircraft in the area—but there were none. Could there have been airplanes in the area that CAA didn't know about? The answer was almost a flat "No." No one would fly 600 miles off the coast without filing a flight plan; if he got into trouble or went down, the Coast Guard or Air Rescue Service would have no idea where to look.

Kerry was given the same negative answer when he checked on surface shipping.

The last possibility was that the UFO's were meteors, but several points in the pilots' story ruled these out. First, there was a solid overcast at about 18,000 feet. No meteor cruises along straight and level below 18,000 feet. Second, on only rare occasions have meteors been seen traveling three in trail. The chances of seeing such a phenomenon are well over one in a billion.

Some people have guessed that some kind of an atmospheric phenomenon can form a "wall of air" ahead of an airplane that will act as a mirror and that lights seen at night by pilots are nothing more than the reflection of the airplane's own lights. This could be true in some cases, but to have a reflection you must have a light to reflect. There are no lights on an airplane that even approach being "ten times the size of a landing light."

What was it? I know a colonel who says it was the same thing that the two Eastern Airlines' pilots, Clarence Chiles and John Whitted, saw near Montgomery, Alabama, on July 24, 1948, and he thinks that Chiles and Whitted saw a spaceship.

Reports for the month of April set an all-time high. These were all reports that came from military installations. In addition, we received possibly two hundred letters reporting UFO's, but we were so busy all we could do was file them for future reference.

In May 1952 I'd been out to George AFB in California investigating a series of sightings and was on my way home. I remember the flight to Dayton because the weather was bad all the way. I didn't want to miss my connecting flight in Chicago, or get grounded, because I had faithfully promised my wife that we would go out to dinner the night that I returned to Dayton. I'd called her from Los Angeles to tell her that I was coming in, and she had found a baby sitter and had dinner reservations. I hadn't been home more than about two days a week for the past three months, and she was looking forward to going out for the evening.

I reached Dayton about midmorning and went right out to the base. When I arrived at the office, my secretary was gone but there was a big note on my desk: "Call Colonel Dunn as soon as you get in."

I called Colonel Dunn; then I called my wife and told her to cancel the baby sitter, cancel the dinner reservations, and pack my other bag. I had to go to Washington.

While I'd been in California, Colonel Dunn had received a call from General Samford's office. It seems that a few nights before, one of the top people in the Central Intelligence Agency was having a lawn party at his home just outside Alexandria, Virginia. A number of notable personages were in attendance and they had seen a flying saucer. The report had been passed down to Air Force intelligence, and due to the quality of the brass involved, it was "suggested" that I get to Washington on the double and talk to the host of the party. I was at his office before 5:00P.M. and got his report.

About ten o'clock in the evening he and two other people were standing near the edge of his yard talking; he happened to be facing south, looking off across the countryside. He digressed a bit from his story to explain that his home is on a hilltop in the country, and when looking south, he had a view of the entire countryside. While he was talking to the two other people he noticed a light approaching from the west. He had assumed it was an airplane and had casually watched it, but when the light got fairly close, the CIA man said that he suddenly realized there wasn't any sound associated with it. If it were an airplane it would have been close enough for him to hear even above the hum of the guests' conversations. He had actually quit talking and was looking at the light when it stopped for an instant and began to climb almost vertically. He said something to the other guests, and they looked up just in time to see the light finish its climb, stop, and level out. They all watched it travel level for a few seconds, then go into a nearly vertical dive, level out, and streak off to the east.

Most everyone at the party had seen the light before it disappeared, and within minutes several friendly arguments as to what it was had developed, I was told. One person thought it was a lighted balloon, and a retired general thought it was an airplane. To settle the arguments, they had made a few telephone calls. I might add that these people were such that the mention of their names on a telephone got quick results. Radar in the Washington area said that there had been no airplanes flying west to east south of Alexandria in the past hour. The weather station at Bolling AFB said that there were no balloons in the area, but as a double check the weather people looked at their records of high-altitude winds. It couldn't have been a balloon because none of the winds up to 65,000 feet were blowing from west to east—and to be

122

able to see a light on a balloon, it has to be well below 65,000 feet; the man from CIA told me that they had even considered the possibility that the UFO was a meteor and that the "jump" had been due to some kind of an atmospheric distortion. But the light had been in sight too long to be a meteor. He added that an army chaplain and two teetotaler guests had also seen the light jump.

There wasn't much left for me to do when I finished talking to the man. He and his guests had already made all of the checks that I'd have made. All I could do was go back to Dayton, write up his report, and stamp it "Unknown."

Back in March, when it had become apparent that the press was reviving its interest in UFO's, I had suggested that Project Blue Book subscribe to a newspaper clipping service. Such a service could provide several things. First, it would show us exactly how much publicity the UFO's were getting and what was being said, and it would give us the feel of the situation. Then it would also provide a lot of data for our files. In many cases the newspapers got reports that didn't go to the Air Force. Newspaper reporters rival any intelligence officer when it comes to digging up facts, and there was always the possibility that they would uncover and print something we'd missed. This was especially true in the few cases of hoaxes that always accompany UFO publicity. Last, it would provide us with material on which to base a study of the effect of newspaper publicity upon the number and type of UFO reports.

Colonel Dunn liked the idea of the clipping service, and it went into effect soon after the first publicity had appeared. Every three or four days we would get an envelope full of clippings. In March the clipping service was sending the clippings to us in letter-sized envelopes. The envelopes were thin—maybe there would be a dozen or so clippings in each one. Then they began to get thicker and thicker, until the people who were doing the clipping switched to using manila envelopes. Then the manila envelopes began to get thicker and thicker. By May we were up to old shoe boxes. The majority of the newspaper stories in the shoe boxes were based on material that had come from ATIC.

All of these inquiries from the press were adding to Blue Book's work load and to my problems. Normally a military unit such as ATIC has its own public information officer, but we had none so I was it. I was being quoted quite freely in the press and was repeatedly being snarled at by someone in the Pentagon. It was almost a daily occurrence to have people from the "puzzle palace" call and indignantly ask, "Why did you tell them that?" They usually referred to some bit of information that somebody didn't think should have been released. I finally gave up and complained to Colonel Dunn. I suggested that any contacts with the press be made through the Office of Public Information in the Pentagon. These people were trained and paid to do this job; I wasn't. Colonel Dunn heartily agreed because every time I got chewed out he at least got a dirty look.

Colonel Dunn called General Samford's office and they brought in General Sory Smith of the Department of Defense, Office of Public Information. General Smith appointed a civilian on the Air Force Press Desk, Al Chop, to handle all inquiries from the press. The plan was that Al would try to get his answers from Major Dewey Fournet, Blue Book's liaison officer in the Pentagon, and if Dewey didn't have the answer, Al had permission to call me.

This arrangement worked out fine because Al Chop had been through previous UFO publicity battles when he was in the Office of Public Information at Wright Field.

The interest in the UFO's that was shown by the press in May was surpassed only by the interest of the Pentagon. Starting in May, I gave on the average of one briefing in Washington every two weeks, and there was always a full house. From the tone of the official comments to the public about UFO's, it would indicate that there wasn't a great deal of interest, but nothing could be further from the truth. People say a lot of things behind a door bearing a sign that reads "Secret Briefing in Progress."

After one of the briefings a colonel (who is now a brigadier general) presented a plan that called for using several flights of F- 94C jet interceptors for the specific purpose of trying to get some good photographs of UFO's. The flight that he proposed would be an operational unit with six aircraft—two would be on constant alert. The F-94C's, then the hottest operational jet we had, would be stripped of all combat gear to give them peak performance, and they would carry a special camera in the nose. The squadrons would be located at places in the United States where UFO's were most frequently seen.

The plan progressed to the point of estimating how soon enough airplanes for two flights could be stripped, how soon special cameras could be built, and whether or not two specific Air Force bases in the U.S. could support the units.

Finally the colonel's plan was shelved, but not because he was considered to be crazy. After considerable study and debate at high command level, it was decided that twelve F-94C's couldn't be spared for the job and it would have been ineffective to use fewer airplanes.

The consideration that the colonel's plan received was an indication of how some of the military people felt about the importance of finding out exactly what the UFO's really were. And in the discussions the words "interplanetary craft" came up more than once.

Requests for briefings came even from the highest figure in the Air Force, Thomas K. Finletter, then the Secretary for Air. On May 8, 1952, Lieutenant Colonel R. J. Taylor of Colonel Dunn's staff and I presented an hour-long briefing to Secretary Finletter and his staff. He listened intently and asked several questions about specific sightings when the briefing was finished. If he was at all worried about the UFO's he certainly didn't show it. His only comment was, "You're doing a fine job, Captain. It must be interesting. Thank you."

Then he made the following statement for the press:

"No concrete evidence has yet reached us either to prove or disprove the existence of the so-called flying saucers. There remain, however, a number of sightings that the Air Force investigators have been unable to explain. As long as this is true, the Air Force will continue to study flying saucer reports."

In May 1952, Project Blue Book received seventy-nine UFO reports compared to ninety-nine in April. It looked as if we'd passed the peak and were now on the downhill side. The 178 reports of the past two months, not counting the thousand or so letters that we'd received directly from the public, had piled up a sizable backlog since we'd had time to investigate and analyze only the better reports. During June we planned to clear out the backlog, and then we could relax.

But never underestimate the power of a UFO. In June the big flap hit —they began to deliver clippings in big cardboard cartons.

Chapter Eleven - The Big Flap

In early June 1952, Project Blue Book was operating according to the operational plan that had been set up in January 1952. It had taken six months to put the plan into effect, and to a person who has never been indoctrinated into the ways of the military, this may seem like a long time. But consult your nearest government worker and you'll find that it was about par for the red tape course.

We had learned early in the project that about 60 per cent of the reported UFO's were actually balloons, airplanes, or astronomical bodies viewed under unusual conditions, so our operational plan was set up to quickly weed out this type of report. This would give us more time to concentrate on the unknown cases.

To weed out reports in which balloons, airplanes, and astronomical bodies were reported as UFO's, we utilized a flow of data that continually poured into Project Blue Book. We received position reports on all flights of the big skyhook balloons and, by merely picking up the telephone, we could get the details about the flight of any other research balloon or regularly scheduled weather balloon in the United States. The location of aircraft in an area where a UFO had been reported was usually checked by the intelligence officer who made the report, but we double-checked his findings by requesting the location of flights from CAA and military air bases. Astronomical almanacs and journals, star charts, and data that we got from observatories furnished us with clues to UFO's that might be astronomical bodies. All of our investigations in this category of report were double-checked by Project Bear's astronomer.

Then we had our newspaper clipping file, which gave us many clues. Hydrographic bulletins and Notams (notices to airmen), published by the government, sometimes gave us other clues. Every six hours we received a complete set of weather data. A dozen or more other sources of data that might shed some light on a reported UFO were continually being studied.

To get all this information on balloons, aircraft, astronomical bodies, and what have you, I had to co-ordinate Project Blue Book's operational plan with the Air Force's Air Weather Service, Flight Service, Research and Development Command, and Air Defense Command with the Navy's Office of Naval Research, and the aerology branch of the Bureau of Aeronautics; and with the Civil Aeronautics Administration, Bureau of Standards, several astronomical observatories, and our own Project Bear. Our entire operational plan was similar to a Model A Ford I had while I was in high school—just about the time you would get one part working, another part would break down.

When a report came through our screening process and still had the "Unknown" tag on it, it went to the MO file, where we checked its characteristics against other reports. For example, on May 25 we had a report from Randolph AFB, Texas. It went through the screening process and came out "Unknown"; it wasn't a balloon, airplane, or astronomical body. So then it went to the MO file. It was a flock of ducks reflecting the city lights. We knew that the Texas UFO's were ducks because our MO file showed that we had an identical report from Moorhead, Minnesota, and the UFO's at Moorhead were ducks.

Radar reports that came into Blue Book went to the radar specialists of ATIC's electronics branch.

Sifting through reams of data in search of the answers to the many reports that were pouring in each week required many hours of overtime work, but when a report came out with the final conclusion, "Unknown," we were sure that it was unknown.

To operate Project Blue Book, I had four officers, two airmen, and two civilians on my permanent staff. In addition, there were three scientists employed full time on Project Bear, along with several others who worked part time. In the Pentagon, Major Fournet, who had taken on the Blue Book liaison job as an extra duty, was now spending full time on it. If you add to this the number of intelligence officers all over the world who were making preliminary investigations and interviewing UFO observers, Project Blue Book was a sizable effort.

Only the best reports we received could be personally investigated in the field by Project Blue Book personnel. The vast majority of the reports had to be evaluated on the basis of what the intelligence officer who had written the report had been able to uncover, or what data we could get by telephone or by mailing out a questionnaire. Our instructions for "what to do before the Blue Book man arrives," which had been printed in many service publications, were beginning to pay off and the reports were continually getting more detailed.

The questionnaire we were using in June 1952 was the one that had recently been developed by Project Bear. Project Bear, along with psychologists from a midwestern university, had worked on it for five months. Many test models had been tried before it reached its final form—the standard questionnaire that Blue Book is using today.

It ran eight pages and had sixty-eight questions which were booby-trapped in a couple of places to give us a cross check on the reliability of the reporter as an observer. We received quite a few questionnaires answered in such a way that it was obvious that the observer was drawing heavily on his imagination.

From this standard questionnaire the project worked up two more specialized types. One dealt with radar sightings of UFO's, the other with sightings made from airplanes.

In Air Force terminology a "flap" is a condition, or situation, or state of being of a group of people characterized by an advanced degree of confusion that has not quite yet reached panic proportions. It can be brought on by any number of things, including the unexpected visit of an inspecting general, a major administrative reorganization, the arrival of a hot piece of intelligence information, or the dramatic entrance of a well-stacked female into an officers' club bar.

In early June 1952 the Air Force was unknowingly in the initial stages of a flap—a flying saucer flap—*the* flying saucer flap of 1952. The situation had never been duplicated before, and it hasn't been duplicated since. All records for the number of UFO reports were not just broken, they were disintegrated. In 1948, 167 UFO reports had come into ATIC; this was considered a big year. In June 1952 we received 149. During the four years the Air Force had been in the UFO business, 615 reports had been collected. During the "Big Flap" our incoming-message log showed 717 reports.

To anyone who had anything to do with flying saucers, the summer of 1952 was just one big swirl of UFO reports, hurried trips, midnight telephone calls, reports to the Pentagon, press interviews, and very little sleep.

If you can pin down a date that the Big Flap started, it would probably be about June 1.

It was also on June 1 that we received a good report of a UFO that had been picked up on radar. June 1 was a Sunday, but I'd been at the office all day getting ready to go to Los Alamos the next day. About 5:00P.M. the telephone rang and the operator told me that I had a long-distance call from California. My caller was the chief of a radar test section for Hughes Aircraft Company in Los Angeles, and he was very excited about a UFO he had to report.

That morning he and his test crew had been checking out a new late- model radar to get it ready for some tests they planned to run early Monday morning. To see if their set was functioning properly, they had been tracking jets in the Los Angeles area. About midmorning, the Hughes test engineer told me, the jet traffic had begun to drop off, and they were about ready to close down their operation when one of the crew picked up a slow-moving target coming across the San Gabriel Mountains north of Los Angeles. He tracked the target for a few minutes and, from the speed and altitude, decided that it was a DC-3. It was at 11,000 feet and traveling about 180 miles an hour toward Santa Monica. The operator was about ready to yell at the other crew members to shut off the set when he noticed something mighty odd— there was a big gap between the last and the rest of the regularly spaced bright spots on the radarscope. The man on the scope called the rest of the crew in because DC-3's just don't triple their speed. They watched the target as it made a turn and started to climb over Los Angeles. They plotted one, two, three, and then four points during the target's climb; then one of the crew grabbed a slide rule. Whatever it was, it was climbing 35,000 feet per minute and traveling about 550 miles an hour in the process. Then as they watched the scope, the target leveled out for a few seconds, went into a high-speed dive, and again leveled out at 55,000 feet. When they lost the target, it was heading southeast somewhere near Riverside, California.

During the sighting my caller told me that when the UFO was only about ten miles from the radar site two of the crew had gone outside but they couldn't see anything. But, he explained, even the high- flying jets that they had been tracking hadn't been leaving vapor trails.

The first thing I asked when the Hughes test engineer finished his story was if the radar set had been working properly. He said that as soon as the UFO had left the scope they had run every possible check on the radar and it was O.K.

I was just about to ask my caller if the target might not have been some experimental airplane from Edwards AFB when he second-guessed me. He said that after sitting around looking at each other for about a minute, some-one suggested that they call Edwards. They did, and Edwards' flight opera-tions told them that they had nothing in the area.

I asked him about the weather. The target didn't look like a weather target was the answer, but just to be sure, the test crew had checked. One of his men was an electronics-weather specialist whom he had hired because of his knowledge of the idiosyncrasies of radar under certain weather conditions. This man had looked into the weather angle. He had gotten the latest weath-er data and checked it, but there wasn't the slightest indication of an inver-sion or any other weather that would cause a false target.

Just before I hung up I asked the man what he thought he and his crew had picked up, and once again I got the same old answer: "Yesterday at this time any of us would have argued for hours that flying saucers were a bunch of nonsense but now, regardless of what you'll say about what we saw, it was something damned real."

I thanked the man for calling and hung up. We couldn't make any more of an analysis of this report than had already been made, it was another un-known.

I went over to the MO file and pulled out the stack of cards behind the tab "High-Speed Climb." There must have been at least a hundred cards, each one representing a UFO report in which the reported object made a high-speed climb. But this was the first time radar had tracked a UFO during a climb.

During the early part of June, Project Blue Book took another jump up on the organizational chart. A year before the UFO project had consisted of one officer. It had risen from the one-man operation to a project within a group, then to a group, and now it was a section. Neither Project Sign nor the old Project Grudge had been higher than the project-within-a-group level. The chief of a group normally calls for a lieutenant colonel, and since I was just a captain this caused some consternation in the ranks. There was some talk about putting Lieutenant Colonel Ray Taylor of Colonel Dunn's staff in charge. Colonel Taylor was very much interested in UFO's; he had handled some of the press contacts prior to turning this function over to the Pentagon and had gone along with me on briefings, so he knew something about the project. But in the end Colonel Donald Bower, who was my division chief, decided rank be damned, and I stayed on as chief of Project Blue Book.

The location within the organizational chart is always indicative of the importance placed on a project. In June 1952 the Air Force was taking the UFO problem seriously. One of the reasons was that there were a lot of good UFO reports coming in from Korea. Fighter pilots reported seeing silver-colored spheres or disks on several occasions, and radar in Japan, Okinawa, and in Korea had tracked unidentified targets.

In June our situation map, on which we kept a plot of all of our sightings, began to show an ever so slight trend toward reports beginning to bunch up on the east coast. We discussed this build-up, but we couldn't seem to find any explainable reason for it so we decided that we'd better pay special attention to reports coming from the eastern states.

I had this build-up of reports in mind one Sunday night, June 15 to be exact, when the OD at ATIC called me at home and said that we were getting a lot of reports from Virginia. Each report by itself wasn't too good, the OD told me, but together they seemed to mean something. He suggested that I come out and take a look at them—so I did.

Individually they weren't too good, but when I lined them up chronologically and plotted them on a map they took the form of a hot report.

At 3:40P.M. a woman at Unionville, Virginia, had reported a "very shiny object" at high altitude.

At 4:20P.M. the operators of the CAA radio facility at Gordonsville, Virginia, had reported that they saw a "round, shiny object." It was southeast of their station, or directly south of Unionville.

At 4:25P.M. the crew of an airliner northwest of Richmond, Virginia, reported a "silver sphere at eleven o'clock high."

At 4:43P.M. a Marine pilot in a jet tried to intercept a "round shiny sphere" south of Gordonsville.

At 5:43P.M. an Air Force T-33 jet tried to intercept a "shiny sphere" south of Gordonsville. He got above 35,000 feet and the UFO was still far above him.

At 7:35P.M. many people in Blackstone, Virginia, about 80 miles south of Gordonsville, reported it. It was a "round, shiny object with a golden glow" moving from north to south. By this time radio commentators in central Virginia were giving a running account of the UFO's progress.

At 7:59P.M. the people in the CAA radio facility at Blackstone saw it.

At 8:00P.M. jets arrived from Langley AFB to attempt to intercept it, but at 8:05P.M. it disappeared.

This was a good report because it was the first time we ever received a series of reports on the same object, and there was no doubt that all these people had reported the same object. Whatever it was, it wasn't moving too fast, because it had traveled only about 90 miles in four hours and twenty-five minutes. I was about ready to give up until morning and go home when my wife called. The local Associated Press man had called our home and she assumed that it was about this sighting. She had just said that I was out so he might not call the base. I decided that I'd better keep working so I'd have the answer in time to keep the story out of the papers. A report like this could cause some excitement.

The UFO obviously wasn't a planet because it was moving from north to south, and it was too slow to be an airplane. I called the balloon- plotting center at Lowry AFB, where the tracks of the big skyhook balloons are plotted, but the only big balloons in the air were in the western United States, and they were all accounted for.

It might have been a weather balloon. The wind charts showed that the high-altitude winds were blowing in different directions at different altitudes above 35,000 feet, so there was no one flow of air that could have brought a balloon in from a certain area, and I knew that the UFO had to be higher than 35,000 feet because the T-33 jet had been this high and the UFO was still above it. The only thing to do was to check with all of the weather stations in the area. I called Richmond, Roanoke, several places in the vicinity of Washington, D.C., and four or five other weather stations, but all of their balloons were accounted for and none had been anywhere close to the central part of Virginia.

A balloon can travel only so far, so there was no sense in checking stations too far away from where the people had seen the UFO, but I took a chance and called Norfolk; Charleston, West Virginia; Altoona, Pennsylvania; and other stations within a 150-mile radius of Gordonsville and Blackstone. Nothing.

I still thought it might be a balloon, so I started to call more stations. At Pittsburgh I hit a lead. Their radiosonde balloon had gone up to about 60,000 feet and evidently had sprung a slow leak because it had leveled off at that altitude. Normally balloons go up till they burst at 80,000 or 90,000 feet. The weather forecaster at Pittsburgh said that their records showed they had lost contact with the balloon when it was about 60 miles southeast of their station. He said that the winds at 60,000 feet were constant, so it shouldn't be too difficult to figure out where the balloon went after they had lost it. Things

130

must be dull in Pittsburgh at 2:00 a.m. on Monday mornings, because he offered to plot the course that the balloon probably took and call me back.

In about twenty minutes I got my call. It probably was their balloon, the forecaster said. Above 50,000 feet there was a strong flow of air southeast from Pittsburgh, and this fed into a stronger southerly flow that was paralleling the Atlantic coast just east of the Appalachian Mountains. The balloon would have floated along in this flow of air like a log floating down a river. As close as he could estimate, he said, the balloon would arrive in the Gordonsville- Blackstone area in the late afternoon or early evening. This was just about the time the UFO had arrived.

"Probably a balloon" was a good enough answer for me.

The next morning at 8:00A.M., Al Chop called from the Pentagon to tell me that people were crawling all over his desk wanting to know about a sighting in Virginia.

The reports continued to come in. At Walnut Lake, Michigan, a group of people with binoculars watched a "soft white light" go back and forth across the western sky for nearly an hour. A UFO "paced" an Air Force B-25 for thirty minutes in California. Both of these happened on June 18, and although we checked and rechecked them, they came out as unknowns.

On June 19 radar at Goose AFB in Newfoundland picked up some odd targets. The targets came across the scope, suddenly enlarged, and then became smaller again. One unofficial comment was that the object was flat or disk-shaped, and that the radar target had gotten bigger because the disk had banked in flight to present a greater reflecting surface. ATIC's official comment was weather.

Goose AFB was famous for unusual reports. In early UFO history someone had taken a very unusual colored photo of a "split cloud." The photographer had seen a huge ball of fire streak down through the sky and pass through a high layer of stratus clouds. As the fireball passed through the cloud it cut out a perfect swath. The conclusion was that the fireball was a meteor, but the case is still one of the most interesting in the file because of the photograph.

Then in early 1952 there was another good report from this area. It was an unknown.

The incident started when the pilot of an Air Force C-54 transport radioed Goose AFB and said that at 10:42P.M. a large fireball had buzzed his airplane. It had come in from behind the C-54, and nobody had seen it until it was just off the left wing. The fireball was so big that the pilot said it looked as if it was only a few hundred feet away. The C-54 was 200 miles southwest, coming into Goose AFB from Westover AFB, Massachusetts, when the incident occurred. The base officer-of-the-day, who was also a pilot, happened to be in the flight operations office at Goose when the message came in and he overheard the report. He stepped outside, walked over to his command car, and told his driver about the radio message, so the driver got out and both of them looked toward the south. They searched the horizon for a few seconds; then suddenly they saw a light closing in from the southwest. Within a second, it was near the airfield. It had increased in size till it was as big as a "golf ball at arm's length," and it looked like a big ball of fire. It was so low that both the OD and his driver dove under the command car because they were sure it was going to hit the airfield. When they turned and looked up they saw the fireball make a 90-degree turn over the airfield and disappear into the northwest. The time was 10:47P.M.

The control tower operators saw the fireball too, but didn't agree with the OD and his driver on how low it was. They did think that it had made a 90-degree turn and they didn't think that it was a meteor. In the years they'd been in towers they'd seen hundreds of meteors, but they'd never seen anything like this, they reported.

And reports continued to pour into Project Blue Book. It was now not uncommon to get ten or eleven wires in one day. If the letters reporting UFO sightings were counted, the total would rise to twenty or thirty a day. The majority of the reports that came in by wire could be classified as being good. They were reports made by reliable people and they were full of details. Some were reports of balloons, airplanes, etc., but the percentage of unknowns hovered right around 22 per cent.

To describe and analyze each report, or even the unknowns, would require a book the size of an unabridged dictionary, so I am covering only the best and most representative cases.

One day in mid-June, Colonel Dunn called me. He was leaving for Washington and he wanted me to come in the next day to give a briefing at a meeting. By this time I was taking these briefings as a matter of course. We usually gave the briefings to General Garland and a general from the Research and Development Board, who passed the information on to General Samford, the Director of Intelligence. But this time General Samford, some of the members of his staff, two Navy captains from the Office of Naval Intelligence, and some people I can't name were at the briefing.

When I arrived in Washington, Major Fournet told me that the purpose of the meetings, and my briefing, was to try to find out if there was any significance to the almost alarming increase in UFO reports over the past few weeks. By the time that everyone had finished signing into the briefing room in the restricted area of the fourth-floor "B" ring of the Pentagon, it was about 9:15A.M. I started my briefing as soon as everyone was seated.

I reviewed the last month's UFO activities; then I briefly went over the more outstanding "Unknown" UFO reports and pointed out how they were increasing in number—breaking all previous records. I also pointed out that even though the UFO subject was getting a lot of publicity, it wasn't the scare-type publicity that had accompanied the earlier flaps—in fact, much of the present publicity was anti- saucer.

Then I went on to say that even though the reports we were getting were detailed and contained a great deal of good data, we still had no proof the UFO's were anything real. We could, I said, prove that all UFO reports were

132

merely the misinterpretation of known objects *if* we made a few assumptions.

At this point one of the colonels on General Samford's staff stopped me. "Isn't it true," he asked, "that if you make a few positive assumptions instead of negative assumptions you can just as easily prove that the UFO's are interplanetary spaceships? Why, when you have to make an assumption to get an answer to a report, do you always pick the assumption that proves the UFO's don't exist?"

You could almost hear the colonel add, "O.K., so now I've said it."

For several months the belief that Project Blue Book was taking a negative attitude and the fact that the UFO's could be interplanetary spaceships had been growing in the Pentagon, but these ideas were usually discussed only in the privacy of offices with doors that would close tight.

No one said anything, so the colonel who had broken the ice plunged in. He used the sighting from Goose AFB, where the fireball had buzzed the C-54 and sent the OD and his driver belly-whopping under the command car as an example. The colonel pointed out that even though we had labeled the report "Unknown" it wasn't accepted as proof. He wanted to know why.

I said that our philosophy was that the fireball could have been two meteors: one that buzzed the C-54 and another that streaked across the airfield at Goose AFB. Granted a meteor doesn't come within feet of an airplane or make a 90-degree turn, but these could have been optical illusions of some kind. The crew of the C-54, the OD, his driver, and the tower operators didn't recognize the UFO's as meteors because they were used to seeing the normal "shooting stars" that are most commonly seen.

But the colonel had some more questions. "What are the chances of having two extremely spectacular meteors in the same area, traveling the same direction, only five minutes apart?"

I didn't know the exact mathematical probability, but it was rather small, I had to admit.

Then he asked, "What kind of an optical illusion would cause a meteor to appear to make a 90-degree turn?"

I had asked our Project Bear astronomer this same question, and he couldn't answer it either. So the only answer I could give the colonel was, "I don't know." I felt as if I were on a witness stand being cross-examined, and that is exactly where I was, because the colonel cut loose.

"Why not assume a point that is more easily proved?" he asked. "Why not assume that the C-54 crew, the OD, his driver, and the tower operators did know what they were talking about? Maybe they had seen spectacular meteors during the hundreds of hours that they had flown at night and the many nights that they had been on duty in the tower. Maybe the ball of fire had made a 90-degree turn. Maybe it was some kind of an intelligently controlled craft that had streaked northeast across the Gulf of St. Lawrence and Quebec Province at 2,400 miles an hour.

"Why not just simply believe that most people know what they saw?" the colonel said with no small amount of sarcasm in his voice.

This last comment started a lively discussion, and I was able to retreat. The colonel had been right in a sense—we were being conservative, but maybe this was the right way to be. In any scientific investigation you always assume that you don't have enough proof until you get a positive answer. I don't think that we had a positive answer—yet.

The colonel's comments split the group, and a hot exchange of ideas, pros and cons, and insinuations that some people were imitating ostriches to keep from facing the truth followed.

The outcome of the meeting was a directive to take further steps to obtain positive identification of the UFO's. Our original idea of attempting to get several separate reports from one sighting so we could use triangulation to measure speed, altitude, and size wasn't working out. We had given the idea enough publicity, but reports where triangulation could be used were few and far between. Mr. or Mrs. Average Citizen just doesn't look up at the sky unless he or she sees a flash of light or hears a sound. Then even if he or she does look up and sees a UFO, it is very seldom that the report ever gets to Project Blue Book. I think that it would be safe to say that Blue Book only heard about 10 per cent of the UFO's that were seen in the United States.

After the meeting I went back to ATIC, and the next day Colonel Don Bower and I left for the west coast to talk to some people about how to get better UFO data. We brought back the idea of using an extremely long focal-length camera equipped with a diffraction grating.

The cameras would be placed at various locations throughout the United States where UFO's were most frequently seen. We hoped that photos of the UFO's taken through the diffraction gratings would give us some proof one way or the other.

The diffraction gratings we planned to use over the lenses of the cameras were the same thing as prisms; they would split up the light from the UFO into its component parts so that we could study it and determine whether it was a meteor, an airplane, or balloon reflecting sunlight, etc. Or we might be able to prove that the photographed UFO was a craft completely foreign to our knowledge.

A red-hot, A-1 priority was placed on the camera project, and a section at ATIC that developed special equipment took over the job of obtaining the cameras, or, if necessary, having them designed and built.

But the UFO's weren't waiting around till they could be photographed. Every day the tempo and confusion were increasing a little more.

By the end of June it was very noticeable that most of the better reports were coming from the eastern United States. In Massachusetts, New Jersey, and Maryland jet fighters had been scrambled almost nightly for a week. On three occasions radar-equipped F-94's had locked on aerial targets only to have the lock-on broken by the apparent violent maneuvers of the target.

By the end of June there was also a lull in the newspaper publicity about the UFO's. The forthcoming political conventions had wiped out any mention of flying saucers. But on July 1 there was a sudden outbreak of good reports. The first one came from Boston; then they worked down the coast.

About seven twenty-five on the morning of July 1 two F-94's were scrambled to intercept a UFO that a Ground Observer Corps spotter reported was traveling southwest across Boston. Radar couldn't pick it up so the two airplanes were just vectored into the general area. The F-94's searched the area but couldn't see anything. We got the report at ATIC and would have tossed it out if it hadn't been for other reports from the Boston area at that same time.

One of these reports came from a man and his wife at Lynn, Massachusetts, nine miles northeast of Boston. At seven-thirty they had noticed the two vapor trails from the climbing jet interceptors. They looked around the sky to find out if they could see what the jets were after and off to the west they saw a bright silver "cigar- shaped object about six times as long as it was wide" traveling southwest across Boston. It appeared to be traveling just a little faster than the two jets. As they watched they saw that an identical UFO was following the first one some distance back. The UFO's weren't leaving vapor trails but, as the man mentioned in his report, this didn't mean anything because you can get above the vapor trail level. And the two UFO's appeared to be at a very high altitude. The two observers watched as the two F-94's searched back and forth far below the UFO's.

Then there was another report, also made at seven-thirty. An Air Force captain was just leaving his home in Bedford, about 15 miles northwest of Boston and straight west of Lynn, when he saw the two jets. In his report he said that he, too, had looked around the sky to see if he could see what they were trying to intercept when off to the east he saw a "silvery cigar-shaped object" traveling south. His description of what he observed was almost identical to what the couple in Lynn reported except that he saw only one UFO.

When we received the report, I wanted to send someone up to Boston immediately in the hope of getting more data from the civilian couple and the Air Force captain; this seemed to be a tailor-made case for triangulation. But by July 1 we were completely snowed under with reports, and there just wasn't anybody to send. Then, to complicate matters, other reports came in later in the day.

Just two hours after the sighting in the Boston area Fort Monmouth, New Jersey, popped back into UFO history. At nine-thirty in the morning twelve student radar operators and three instructors were tracking nine jets on an SCR 584 radar set when two UFO targets appeared on the scope. The two targets came in from the northeast at a slow speed, much slower than the jets that were being tracked, hovered near Fort Monmouth at 50,000 feet for about five minutes, and then took off in a "terrific burst of speed" to the southwest.

When the targets first appeared, some of the class went outside with an instructor, and after searching the sky for about a minute, they saw two shiny objects in the same location as the radar showed the two unidentified targets to be. They watched the two UFO's for several minutes and saw them go zipping off to the southwest at exactly the same time that the two radar targets moved off the scope in that direction.

We had plotted these reports, the ones from Boston and the one from Fort Monmouth, on a map, and without injecting any imagination or wild assumptions, it looked as if two "somethings" had come down across Boston on a southwesterly heading, crossed Long Island, hovered for a few minutes over the Army's secret laboratories at Fort Monmouth, then proceeded toward Washington. In a way we half expected to get a report from Washington. Our expectations were rewarded because in a few hours a report arrived from that city.

A physics professor at George Washington University reported a "dull, gray, smoky-colored" object which hovered north northwest of Washington for about eight minutes. Every once in a while, the professor reported, it would move through an arc of about 15 degrees to the right or left, but it always returned to its original position. While he was watching the UFO he took a 25-cent piece out of his pocket and held it at arm's length so that he could compare its size to that of the UFO. The UFO was about half the diameter of the quarter. When he first saw the UFO, it was about 30 to 40 degrees above the horizon, but during the eight minutes it was in sight it steadily dropped lower and lower until buildings in downtown Washington blocked off the view.

Besides being an "Unknown," this report was exceptionally interesting to us because the sighting was made from the center of downtown Washington, D.C. The professor reported that he had noticed the UFO when he saw people all along the street looking up in the air and pointing. He estimated that at least 500 people were looking at it, yet his was the only report we received. This seemed to substantiate our theory that people are very hesitant to report UFO's to the Air Force. But they evidently do tell the newspapers because later on we picked up a short account of the sighting in the Washington papers. It merely said that hundreds of calls had been received from people reporting a UFO.

When reports were pouring in at the rate of twenty or thirty a day, we were glad that people were hesitant to report UFO's, but when we were trying to find the answer to a really knotty sighting we always wished that more people had reported it. The old adage of having your cake and eating it, too, held even for the UFO.

Technically no one in Washington, besides, of course, Major General Samford and his superiors, had anything to do with making policy decisions about the operation of Project Blue Book or the handling of the UFO situation in general. Nevertheless, everyone was trying to get into the act. The split in opinions on what to do about the rising tide of UFO reports, the split that first came out in the open at General Samford's briefing, was widening every day. One group was getting dead-serious about the situation. They thought we now had plenty of evidence to back up an official statement that the UFO's were something real and, to be specific, not something from this earth. This group wanted Project Blue Book to quit spending time investigating reports from the standpoint of trying to determine if the observer of a UFO had actually seen something foreign to our knowledge and start assuming that he or she had. They wanted me to aim my investigation at trying to find out more about the UFO. Along with this switch in operating policy, they wanted to clamp down on the release of information. They thought that the security classification of the project should go up to Top Secret until we had all of the answers, then the information should be released to the public. The investigation of UFO's along these lines should be a maximum effort, they thought, and their plans called for lining up many top scientists to devote their full time to the project. Someone once said that enthusiasm is infectious, and he was right. The enthusiasm of this group took a firm hold in the Pentagon, at Air Defense Command Headquarters, on the Research and Development Board, and many other agencies throughout the government. But General Samford was still giving the orders, and he said to continue to operate just as we had—keeping an open mind to any ideas.

After the minor flurry of reports on July 1 we had a short breathing spell and found time to clean up a sizable backlog of reports. People were still seeing UFO's but the frequency of the sighting curve was dropping steadily. During the first few days of July we were getting only two or three good reports a day.

On July 5 the crew of a non-scheduled airliner made page two of many newspapers by reporting a UFO over the AEC's supersecret Hanford, Washington, installation. It was a skyhook balloon. On the twelfth a huge meteor sliced across Indiana, southern Illinois, and Missouri that netted us twenty or thirty reports. Even before they had stopped coming in, we had confirmation from our astronomer that the UFO was a meteor.

But forty-two minutes later there was a sighting in Chicago that wasn't so easily explained.

According to our weather records, on the night of July 12 it was hot in Chicago. At nine forty-two there were at least 400 people at Montrose Beach trying to beat the heat. Many of them were lying down looking at the stars, so that they saw the UFO as it came in from the west northwest, made a 180-degree turn directly over their heads, and disappeared over the horizon. It was a "large red light with small white lights on the side," most of the people reported. Some of them said that it changed to a single yellow light as it made its turn. It was in sight about five minutes, and during this time no one reported hearing any sound.

One of the people at the beach was the weather officer from O'Hare International Airport, an Air Force captain. He immediately called O'Hare. They checked on balloon flights and with radar, but both were negative; radar said that there had been no aircraft in the area of Montrose Beach for several hours.

I sent an investigator to Chicago, and although he came back with a lot of data on the sighting, it didn't add up to be anything known.

The next day Dayton had its first UFO sighting in a long time when a Mr. Roy T. Ellis, president of the Rubber Seal Products Company, and many other people, reported a teardrop-shaped object that hovered over Dayton for several minutes about midnight. This sighting had an interesting twist because two years later I was in Dayton and stopped in at ATIC to see a friend who is one of the technical advisers at the center.

Naturally the conversation got around to the subject of UFO's, and he asked me if I remembered this specific sighting. I did, so he went on to say that he and his wife had seen this UFO that night but they had never told anybody. He was very serious when he admitted that he had no idea what it could have been. Now I'd heard this statement a thousand times before from other people, but coming from this person, it was really something because he was as anti-saucer as anyone I knew. Then he added, "From that time on I didn't think your saucer reporters were as crazy as I used to think they were."

The Dayton sighting also created quite a stir in the press. In conjunction with the sighting, the Dayton Daily *Journal* had interviewed Colonel Richard H. Magee, the Dayton-Oakwood civil defense director; they wanted to know what he thought about the UFO's. The colonel's answer made news: "There's something flying around in our skies and we wish we knew what it was."

When the story broke in other papers, the colonel's affiliation with civil defense wasn't mentioned, and he became merely "a colonel from Dayton." Dayton was quickly construed by the public to mean Wright- Patterson AFB and specifically ATIC. Some people in the Pentagon screamed while others gleefully clapped their hands. The gleeful handclaps were from those people who wanted the UFO's to be socially recognized, and they believed that if they couldn't talk their ideas into being they might be able to force them in with the help of this type of publicity.

The temporary lull in reporting that Project Blue Book had experienced in early July proved to be only the calm before the storm. By mid-July we were getting about twenty reports a day plus frantic calls from intelligence officers all over the United States as every Air Force installation in the U.S. was being swamped with reports. We told the intelligence officers to send in the ones that sounded the best.

The build-up in UFO reports wasn't limited to the United States— every day we would receive reports from our air attaches in other countries. England and France led the field, with the South American countries running a close third. Needless to say, we didn't investigate or evaluate foreign reports because we had our hands full right at home.

Most of us were putting in fourteen hours a day, six days a week. It wasn't at all uncommon for Lieutenant Andy Flues, Bob Olsson, or Kerry Rothstien, my investigators, to get their sleep on an airliner going out or coming back from an investigation. TWA airliners out of Dayton were more like home than home. But we hadn't seen anything yet.

All the reports that were coming in were good ones, ones with no answers. Unknowns were running about 40 percent. Rumors persist that in mid-July 1952 the Air Force was braced for an expected invasion by flying saucers. Had these rumormongers been at ATIC in mid-July they would have thought that the invasion was already in full swing. And they would have thought that one of the beachheads for the invasion was Patrick AFB, the Air Force's Guided Missile Long-Range Proving Ground on the east coast of Florida.

On the night of July 18, at ten forty-five, two officers were standing in front of base operations at Patrick when they noticed a light at about a 45-degree angle from the horizon and off to the west. It was an amber color and "quite a bit brighter than a star." Both officers had heard flying saucer stories, and both thought the light was a balloon. But, to be comedians, they called to several more officers and airmen inside the operations office and told them to come out and "see the flying saucer." The people came out and looked. A few were surprised and took the mysterious light seriously, at the expense of considerable laughter from the rest of the group. The discussion about the light grew livelier and bets that it was a balloon were placed. In the meantime the light had drifted over the base, had stopped for about a minute, turned, and was now heading north. To settle the bet, one of the officers stepped into the base weather office to find out about the balloon. Yes, one was in the air and being tracked by radar, he was told. The weather officer said that he would call to find out exactly where it was. He called and found out that the weather balloon was being tracked due west of the base and that the light had gone out about ten minutes before. The officer went back outside to find that what was first thought to be a balloon was now straight north of the field and still lighted. To add to the confusion, a second amber light had appeared in the west about 20 degrees lower than where the first one was initially seen, and it was also heading north but at a much greater speed. In a few seconds the first light stopped and started moving back south over the base.

While the group of officers and airmen were watching the two lights, the people from the weather office came out to tell the UFO observers that the balloon was still traveling straight west. They were just in time to see a third light come tearing across the sky, directly overhead, from west to east. A weatherman went inside and called the balloon-tracking crew again—their balloon was still far to the west of the base.

Inside of fifteen minutes two more amber lights came in from the west, crossed the base, made a 180-degree turn over the ocean, and came back over the observers.

In the midst of the melee a radar set had been turned on but it couldn't pick up any targets. This did, however, eliminate the possibility of the lights' being aircraft. They weren't stray balloons either, because the winds at all altitudes were blowing in a westerly direction. They obviously weren't meteors. They weren't searchlights on a haze layer because there was no weather conducive to forming a haze layer and there were no searchlights.

They could have been some type of natural phenomenon, if one desires to take the negative approach. Or, if you take the positive approach, they could have been spaceships.

The next night radar at Washington National Airport picked up UFO's and one of the most highly publicized sightings of UFO history was in the making. It marked the beginning of the end of the Big Flap.

Chapter Twelve - The Washington Merry-Go-Round

No flying saucer report in the history of the UFO ever won more world acclaim than the Washington National Sightings.

When radars at the Washington National Airport and at Andrews AFB, both close to the nation's capital, picked up UFO's, the sightings beat the Democratic National Convention out of headline space. They created such a furor that I had inquiries from the office of the President of the United States and from the press in London, Ottawa, and Mexico City. A junior-sized riot was only narrowly averted in the lobby of the Roger Smith Hotel in Washington when I refused to tell U.S. newspaper reporters what I knew about the sightings.

Besides being the most highly publicized UFO sightings in the Air Force annals, they were also the most monumentally fouled-up messes that repose in the files. Although the Air Force said that the incident had been fully investigated, the Civil Aeronautics Authority wrote a formal report on the sightings, and numerous magazine writers studied them, the complete story has never fully been told. The pros have been left out of the con accounts, and the cons were neatly overlooked by the pro writers.

For a year after the twin sightings we were still putting little pieces in the puzzle.

In some aspects the Washington National Sightings could be classed as a surprise—we used this as an excuse when things got fouled up— but in other ways they weren't. A few days prior to the incident a scientist, from an agency that I can't name, and I were talking about the build-up of reports along the east coast of the United States. We talked for about two hours, and I was ready to leave when he said that he had one last comment to make—a prediction. From his study of the UFO reports that he was getting from Air Force Headquarters, and from discussions with his colleagues, he said that he thought that we were sitting right on top of a big keg full of loaded flying saucers. "Within the next few days," he told me, and I remember that he punctuated his slow, deliberate remarks by hitting the desk with his fist, "they're going to blow up and you're going to have the granddaddy of all UFO sightings. The sighting will occur in Washington or New York," he predicted, "probably Washington."

The trend in the UFO reports that this scientist based his prediction on hadn't gone unnoticed. We on Project Blue Book had seen it, and so had the people in the Pentagon; we all had talked about it.

On July 10 the crew of a National Airlines plane reported a light "too bright to be a lighted balloon and too slow to be a big meteor" while they were flying south at 2,000 feet near Quantico, Virginia, just south of Washington.

On July 13 another airliner crew reported that when they were 60 miles southwest of Washington, at 11,000 feet, they saw a light below them. It came up to their level, hovered off to the left for several minutes, and then it took off in a fast, steep climb when the pilot turned on his landing lights.

On July 14 the crew of a Pan American airliner en route from New York to Miami reported eight UFO's near Newport News, Virginia, about 130 miles south of Washington.

Two nights later there was another sighting in exactly the same area but from the ground. At 9:00P.M. a high-ranking civilian scientist from the National Advisory Committee for Aeronautics Laboratory at Langley AFB and another man were standing near the ocean looking south over Hampton Roads when they saw two amber-colored lights, "much too large to be aircraft lights," off to their right, silently traveling north. Just before the two lights got abreast of the two men they made a 180-degree turn and started back toward the spot where they had first been seen. As they turned, the two lights seemed to "jockey for position in the formation." About this time a third light came out of the west and joined the first two; then as the three UFO's climbed out of the area toward the south, several more lights joined the formation. The entire episode had lasted only three minutes.

The only possible solution to the sighting was that the two men had seen airplanes. We investigated this report and found that there were several B-26's from Langley AFB in the area at the time of the sighting, but none of the B-26 pilots remembered being over Hampton Roads. In fact, all of them had generally stayed well south of Norfolk until about 10:30P.M. because of thunderstorm activity northwest of Langley. Then there were other factors—the observers heard no sound and they were away from all city noises, aircraft don't carry just one or two amber lights, and the distance between the two lights was such that had they been on an airplane the airplane would have been huge or very close to the observers. And last, but not least, the man from the National Advisory Committee for Aeronautics was a very famous aerodynamicist and of such professional stature that if he said the lights weren't airplanes they weren't.

This then was the big build-up to the first Washington national sighting and the reason why my friend predicted that the Air Force was sitting on a big powder keg of loaded flying saucers.

When the keg blew the best laid schemes of the mice and men at ATIC, they went the way best laid schemes are supposed to. The first one of the highly publicized Washington national sightings started, according to the CAA's logbook at the airport, at 11:40P.M. on the night of July 19 when two radars at National Airport picked up eight unidentified targets east and south of Andrews AFB. The targets weren't airplanes because they would loaf along at 100 to 130 miles an hour then suddenly accelerate to "fantastically high speeds" and leave the area. During the night the crews of several airliners saw mysterious lights in the same locations that the radars showed the targets; tower operators also saw lights, and jet fighters were brought in.

But nobody bothered to tell Air Force Intelligence about the sighting. When reporters began to call intelligence and ask about the big sighting behind the headlines, INTERCEPTORS CHASE FLYING SAUCERS OVER WASHINGTON, D.C., they were told that no one had ever heard of such a sighting. In the next edition the headlines were supplemented by, AIR FORCE WONT TALK.

Thus intelligence was notified about the first Washington national sighting.

I heard about the sighting about ten o'clock Monday morning when Colonel Donald Bower and I got off an airliner from Dayton and I bought a newspaper in the lobby of the Washington National Airport Terminal Building. I called the Pentagon from the airport and talked to Major Dewey Fournet, but all he knew was what he'd read in the papers. He told me that he had called the intelligence officer at Bolling AFB and that he was making an investigation. We would get a preliminary official report by noon.

It was about 1:00P.M. when Major Fournet called me and said that the intelligence officer from Bolling was in his office with the preliminary report on the sightings. I found Colonel Bower, we went up to Major Fournet's office and listened to the intelligence officer's briefing.

The officer started by telling us about the location of the radars involved in the incident. Washington National Airport, which is located about three miles south of the heart of the city, had two radars. One was a long-range radar in the Air Route Traffic Control section. This radar had 100-mile range and was used to control all air traffic approaching Washington. It was known as the ARTC radar. The control tower at National Airport had a shorter-range radar that it used to control aircraft in the immediate vicinity of the airport. Bolling AFB, he said, was located just east of National Airport, across the Potomac River. Ten miles farther east, in almost a direct line with National and Bolling, was Andrews AFB. It also had a short- range radar. All of these airfields were linked together by an intercom system.

Then the intelligence officer went on to tell about the sighting.

When a new shift took over at the ARTC radar room at National Airport, the air traffic was light so only one man was watching the radarscope. The senior traffic controller and the six other traffic controllers on the shift were out of the room at eleven-forty, when the man watching the radarscope noticed a group of seven targets appear. From their position on the scope he knew that they were just east and a little south of Andrews AFB. In a way the targets looked like a formation of slow airplanes, but no formations were due in the area. As he watched, the targets loafed along at 100 to 130 miles an hour; then in an apparent sudden burst of speed two of them streaked out of radar range. These were no airplanes, the man thought, so he let out a yell for the senior controller. The senior controller took one look at the scope and called in two more of the men. They all agreed that these were no airplanes. The targets could be caused by a malfunction in the radar, they thought, so a technician was called in —the set was in perfect working order.

The senior controller then called the control tower at National Airport; they reported that they also had unidentified targets on their scopes, so did Andrews. And both of the other radars reported the same slow speeds followed by a sudden burst of speed. One target was clocked at 7,000 miles an hour. By now the targets had moved into every sector of the scope and had flown through the prohibited flying areas over the White House and the Capitol.

Several times during the night the targets passed close to commercial airliners in the area and on two occasions the pilots of the airliners saw lights that they couldn't identify, and the lights were in the same spots where the radar showed UFO's to be. Other pilots to whom the ARTC radar men talked on the radio didn't see anything odd, at least that's what they said, but the senior controller knew airline pilots and knew that they were very reluctant to report UFO's.

The first sighting of a light by an airline pilot took place shortly after midnight, when an ARTC controller called the pilot of a Capital Airlines flight just taking off from National. The controller asked the pilot to keep watch for unusual lights—or anything. Soon after the pilot cleared the traffic pattern, and while ARTC was still in contact with him, he suddenly yelled, "There's one— off to the right— and there it goes." The controller had been watching the scope, and a target that had been off to the right of the Capitaliner was gone.

During the next fourteen minutes this pilot reported six more identical lights.

About two hours later another pilot, approaching National Airport from the south, excitedly called the control tower to report that a light was following him at "eight o'clock level." The tower checked their radar-scope and there was a target behind and to the left of the airliner. The ARTC radar also had the airliner and the UFO target. The UFO tagged along behind and to the left of the airliner until it was within four miles of touchdown on the runway. When the pilot reported the light was leaving, the two radarscopes showed that the target was pulling away from the airliner.

Once during the night all three radars, the two at Washington and the one at Andrews AFB, picked up a target three miles north of the Riverdale Radio beacon, north of Washington. For thirty seconds the three radar operators compared notes about the target over the intercom, then suddenly the target was gone—and it left all three radarscopes simultaneously.

But the clincher came in the wee hours of the morning, when an ARTC traffic controller called the control tower at Andrews AFB and told the tower operators that ARTC had a target just south of their tower, directly over the Andrews Radio range station. The tower operators looked and there was a "huge fiery-orange sphere" hovering in the sky directly over their range station.

Not too long after this excitement had started, in fact just after the technician had checked the radar and found that the targets weren't caused by a radar malfunction, ARTC had called for Air Force interceptors to come in and look around. But they didn't show, and finally ARTC called again—then again. Finally, just about daylight, an F-94 arrived, but by that time the targets were gone. The F-94 crew searched the area for a few minutes but they couldn't find anything unusual so they returned to their base.

So ended phase one of the Washington National Sightings.

The Bolling AFB intelligence officer said he would write up the complete report and forward it to ATIC.

That afternoon things bustled in the Pentagon. Down on the first floor Al Chop was doing his best to stave off the press while up on the fourth floor intelligence officers were holding some serious conferences. There was talk of temperature inversions and the false targets they could cause; but the consensus was that a good radar operator could spot inversion-caused targets, and the traffic controllers who operated the radar at Washington National Airport weren't just out of radar school. Every day the lives of thousands of people depended upon their interpretation of the radar targets they saw on their scopes. And you don't get a job like this unless you've spent a good many years watching a luminous line paint targets on a good many radarscopes. Targets caused by inversions aren't rare—in the years that these men had been working with radar they had undoubtedly seen every kind of target, real or false, that radar can detect. They had told the Bolling AFB intelligence officer that the targets they saw were caused by the radar waves' bouncing off a hard, solid object. The Air Force radar operator at Andrews backed them up; so did two veteran airline pilots who saw lights right where the radar showed a UFO to be.

Then on top of all this there were the reports from the Washington area during the previous two weeks—all good—all from airline pilots or equally reliable people.

To say the least, the sighting at Washington National was a jolt.

Besides trying to figure out what the Washington National UFO's were, we had the problem of what to tell the press. They were now beginning to put on a squeeze by threatening to call a congressman— and nothing chills blood faster in the military. They wanted some kind of an official statement and they wanted it soon. Some people in intelligence wanted to say just, "We don't know," but others held out for a more thorough investigation. I happened to be in this latter category. Many times in the past I had seen what first seemed to be a good UFO report completely fall apart under a thorough investigation. I was for stalling the press and working all night if necessary to go into every aspect of the sighting. But to go along with the theme of the Washington National Sightings—confusion—there was a lot of talk but no action and the afternoon passed with no further investigation.

Finally about 4:00P.M. it was decided that the press, who still wanted an official comment, would get an official "No comment" and that I would stay in Washington and make a more detailed investigation.

I called Lieutenant Andy Flues, who was in charge of Project Blue Book while I was gone, to tell him that I was staying over and I found out that they were in a de luxe flap back in Dayton. Reports were pouring out of the teletype machines at the rate of thirty a day and many were as good, if not better, than the Washington incident. I talked this over with Colonel Bower and we decided that even though things were popping back at ATIC the Washington sighting, from the standpoint of national interest, was more important.

Feeling like a national martyr because I planned to work all night if necessary, I laid the course of my investigation. I would go to Washington National Airport, Andrews AFB, airlines offices, the weather bureau, and a half dozen other places scattered all over the capital city. I called the transportation section at the Pentagon to get a staff car but it took me only seconds to find out that the regulations said no staff cars except for senior colonels or generals. Colonel Bower tried—same thing. General Samford and General Garland were gone, so I couldn't get them to try to pressure a staff car out of the hillbilly who was dispatching vehicles. I went down to the finance office—could I rent a car and charge it as travel expense? No—city buses are available. But I didn't know the bus system and it would take me hours to get to all the places I had to visit, I pleaded. You can take a cab if you want to pay for it out of your per diem was the answer. Nine dollars a day per diem and I should pay for a hotel room, meals, and taxi fares all over the District of Columbia. Besides, the lady in finance told me, my travel orders to Washington covered only a visit to the Pentagon. In addition, she said, I was supposed to be on my way back to Dayton right now, and if I didn't go through all the red tape of getting the orders amended I couldn't collect any per diem and technically I'd be AWOL. I couldn't talk to the finance officer, the lady informed me, because he always left at 4:30 to avoid the traffic and it was now exactly five o'clock and she was quitting.

At five-one I decided that if saucers were buzzing Pennsylvania Avenue in formation I couldn't care less. I called Colonel Bower, explained my troubles, and said that I was through. He concurred, and I caught the next airliner to Dayton.

When I returned I dropped in to see Captain Roy James in the radar branch and told him about the sighting. He said that he thought it sounded as if the radar targets had been caused by weather but since he didn't have the finer details he naturally couldn't make any definite evaluation.

The good UFO reports that Lieutenant Flues had told me about when I called him from Washington had tripled in number before I got around to looking at them. Our daily take had risen to forty a day, and about a third of them were classified as unknowns.

More amber-red fights like those seen on July 18 had been observed over the Guided Missile Long-Range Proving Ground at Patrick AFB, Florida. In

145

Uvalde, Texas, a UFO described as "a large, round, silver object that spun on its vertical axis" was seen to cross 100 degrees of afternoon sky in forty-eight seconds. During part of its flight it passed between two towering cumulus clouds. At Los Alamos and Holyoke, Massachusetts, jets had chased UFO's. In both cases the UFO's had been lost as they turned into the sun.

In two night encounters, one in New Jersey and one in Massachusetts, F-94's tried unsuccessfully to intercept unidentified lights reported by the Ground Observer Corps. In both cases the pilots of the radar- nosed jet interceptors saw a light; they closed in and their radar operators got a lock-on. But the lock-ons were broken in a few seconds, in both cases, as the light apparently took violent evasive maneuvers.

Copies of these and other reports were going to the Pentagon, and I was constantly on the phone or having teleconferences with Major Fournet.

When the second Washington National Sighting came along, almost a week to the hour from the first one, by a stroke of luck things weren't too fouled up. The method of reporting the sighting didn't exactly follow the official reporting procedures that are set forth in Air Force Letter 200-5, dated 5 April 1952, Subject: Reporting of Unidentified Flying Objects—but it worked.

I first heard about the sighting about ten o'clock in the evening when I received a telephone call from Bob Ginna, *Life* magazine's UFO expert. He had gotten the word from *Life's* Washington News Bureau and wanted a statement about what the Air Force planned to do. I decided that instead of giving a mysterious "no comment" I would tell the truth: "I have no idea what the Air Force is doing; in all probability it's doing nothing." When he hung up, I called the intelligence duty officer in the Pentagon and I was correct, intelligence hadn't heard about the sighting. I asked the duty officer to call Major Fournet and ask him if he would go out to the airport, which was only two or three miles from his home. When he got the call from the duty officer Major Fournet called Lieutenant Holcomb; they drove to the ARTC radar room at National Airport and found Al Chop already there. So at this performance the UFO's had an official audience; Al Chop, Major Dewey Fournet, and Lieutenant Holcomb, a Navy electronics specialist assigned to the Air Force Directorate of Intelligence, all saw the radar targets and heard the radio conversations as jets tried to intercept the UFO's.

Being in Dayton, 380 miles away, there wasn't much that I could do, but I did call Captain Roy James thinking possibly he might want to talk on the phone to the people who were watching the UFO's on the radarscopes. But Captain James has a powerful dislike for UFO's— especially on Saturday night.

About five o'clock Sunday morning Major Fournet called and told me the story of the second sighting at Washington National Airport:

About 10:30P.M. on July 26 the same radar operators who had seen the UFO's the week before picked up several of the same slow-moving targets. This time the mysterious craft, if that is what they were, were spread out in an arc around Washington from Herndon, Virginia, to Andrews AFB. This time there was no hesitation in following the targets. The minute they appeared on the big 24-inch radarscope one of the controllers placed a plastic marker representing an unidentified target near each blip on the scope. When all the targets had been carefully marked, one of the controllers called the tower and the radar station at Andrews AFB—they also had the unknown targets.

By 11:30P.M. four or five of the targets were continually being tracked at all times, so once again a call went out for jet interceptors. Once again there was some delay, but by midnight two F- 94's from New Castle County AFB were airborne and headed south. The reporters and photographers were asked to leave the radar room on the pretext that classified radio frequencies and procedures were being used in vectoring the interceptors. All civilian air traffic was cleared out of the area and the jets moved in.

When I later found out that the press had been dismissed on the grounds that the procedures used in an intercept were classified, I knew that this was absurd because any ham radio operator worth his salt could build equipment and listen in on any intercept. The real reason for the press dismissal, I learned, was that not a few people in the radar room were positive that this night would be the big night in UFO history—the night when a pilot would close in on and get a good look at a UFO—and they didn't want the press to be in on it.

But just as the two '94's arrived in the area the targets disappeared from the radarscopes. The two jets were vectored into the areas where the radar had shown the last target plots, but even though the visibility was excellent they could see nothing. The two airplanes stayed around a few minutes more, made a systematic search of the area, but since they still couldn't see anything or pick up anything on their radars they returned to their base.

A few minutes after the F-94's left the Washington area, the unidentified targets were back on the radarscopes in that same area.

What neither Major Fournet nor I knew at this time was that a few minutes after the targets left the radarscopes in Washington people in the area around Langley AFB near Newport News, Virginia, began to call Langley Tower to report that they were looking at weird bright lights that were "rotating and giving off alternating colors." A few minutes after the calls began to come in, the tower operators themselves saw the same or a similar light and they called for an interceptor.

An F-94 in the area was contacted and visually vectored to the light by the tower operators. The F-94 saw the light and started toward it, but suddenly it went out, "like somebody turning off a light bulb." The F-94 crew continued their run and soon got a radar lock-on, but it was broken in a few seconds as the target apparently sped away. The fighter stayed in the area for several more minutes and got two more lock-ons, only to have them also broken after a few seconds.

A few minutes after the F-94 over Newport News had the last lock-on broken, the targets came back on the scopes at Washington National.

With the targets back at Washington the traffic controller again called Air Defense Command, and once again two F-94's roared south toward Washington. This time the targets stayed on the radarscopes when the airplanes arrived.

The controllers vectored the jets toward group after group of targets, but each time, before the jets could get close enough to see anything more than just a light, the targets had sped away. Then one stayed put. The pilot saw a light right where the ARTC radar said a target was located; he cut in the F-94's afterburner and went after it, but just like the light that the F-94 had chased near Langley AFB, this one also disappeared. All during the chase the radar operator in the F-94 was trying to get the target on his set but he had no luck.

After staying in the area about twenty minutes, the jets began to run low on fuel and returned to their base. Minutes later it began to get light, and when the sun came up all the targets were gone.

Early Sunday morning, in an interview with the press, the Korean veteran who piloted the F-94, Lieutenant William Patterson, said:

I tried to make contact with the bogies below 1,000 feet, but they [the radar controllers] vectored us around. I saw several bright lights. I was at my maximum speed, but even then I had no closing speed. I ceased chasing them because I saw no chance of overtaking them. I was vectored into new objects. Later I chased a single bright light which I estimated about 10 miles away. I lost visual contact with it about 2 miles.

When Major Fournet finished telling me about the night's activity, my first question was, "How about the radar targets—could they have been caused by weather?"

I knew that Lieutenant Holcomb was a sharp electronics man and that Major Fournet, although no electronics specialist, was a crackerjack engineer, so their opinion meant a lot.

Dewey said that everybody in the radar room was convinced that the targets were very probably caused by solid metallic objects. There had been weather targets on the scope too, he said, but these were common to the Washington area and the controllers were paying no attention to them.

And this something solid could poke along at 100 miles an hour or outdistance a jet, I thought to myself.

I didn't ask Dewey any more because he'd been up all night and wanted to get to bed.

Monday morning Major Ed Gregory, another intelligence officer at ATIC, and I left for Washington, but our flight was delayed in Dayton so we didn't arrive until late afternoon. On the way through the terminal building to get a cab downtown, I picked up the evening papers. Every headline was about the UFO's:

FIERY OBJECTS OUTRUN JETS OVER CAPITAL—INVESTIGATION VEILED IN SECRECY FOLLOWING VAIN CHASE

JETS ALERTED FOR SAUCERS—INTERCEPTORS CHASE LIGHTS IN D.C. SKIES
EXPERT HERE TO PUSH STUDY AS OBJECTS IN SKIES REPORTED AGAIN

I jokingly commented about wondering who the expert was. In a half hour I found out—I was. When Major Gregory and I walked into the lobby of the Roger Smith Hotel to check in, reporters and photographers rose from the easy chairs and divans like a covey of quail. They wanted my secrets, but I wasn't going to tell nor would I pose for pictures while I wasn't telling anything. Newspaper reporters are a determined lot, but Greg ran interference and we reached the elevator without even a "no comment."

The next day was one of confusion. After the first Washington sighting the prevailing air in the section of the Pentagon's fourth floor, which is occupied by Air Force Intelligence, could be described as excitement, but this day it was confusion. There was a maximum of talk and a minimum of action. Everyone agreed that both sightings should be thoroughly investigated, but nobody did anything. Major Fournet and I spent the entire morning "just leaving" for somewhere to investigate "something." Every time we would start to leave, something more pressing would come up.

About 10:00A.M. the President's air aide, Brigadier General Landry, called intelligence at President Truman's request to find out what was going on. Somehow I got the call. I told General Landry that the radar target could have been caused by weather but that we had no proof.

To add to the already confused situation, new UFO reports were coming in hourly. We kept them quiet mainly because we weren't able to investigate them right away, or even confirm the facts. And we wanted to confirm the facts because some of the reports, even though they were from military sources, were difficult to believe.

Prior to the Washington sightings in only a very few of the many instances in which radar had picked up UFO targets had the targets themselves supposedly been seen visually. Radar experts had continually pointed out this fact to us as an indication that maybe all of the radar targets were caused by freak weather conditions. "If people had just seen a light, or an object, near where the radar showed the UFO target to be, you would have a lot more to worry about," radar technicians had told me many times.

Now people were seeing the same targets that the radars were picking up, and not just at Washington.

On the same night as the second Washington sighting we had a really good report from California. An ADC radar had picked up an unidentified target and an F-94C had been scrambled. The radar vectored the jet interceptor into the target, the radar operator in the '94 locked-on to it, and as the airplane closed in the pilot and RO saw that they were headed directly toward a large, yellowish- orange light. For several minutes they played tag with the UFO. Both the radar on the ground and the radar in the F-94 showed that as soon as the airplane would get almost within gunnery range of the UFO it

149

would suddenly pull away at a terrific speed. Then in a minute or two it would slow down enough to let the F-94 catch it again.

When I talked to the F-94 crew on the phone, the pilot said that they felt as if this were just a big aerial cat-and-mouse game—and they didn't like it—at any moment they thought the cat might have pounced.

Needless to say, this was an unknown.

About midmorning on Tuesday, July 29th, Major General John Samford sent word down that he would hold a press conference that afternoon in an attempt to straighten out the UFO situation with the press.

Donald Keyhoe reports on the press conference and the events leading up to it in detail in his book, *Flying Saucers from Outer Space*. He indicates that before the conference started, General Samford sat behind his big walnut desk in Room 3A138 in the Pentagon and battled with his conscience. Should he tell the public "the real truth"—that our skies are loaded with spaceships? No, the public might panic. The only answer would be to debunk the UFO's.

This bit of reporting makes Major Keyhoe the greatest journalist in history. This beats wire tapping. He reads minds. And not only that, he can read them right through the walls of the Pentagon. But I'm glad that Keyhoe was able to read the General's mind and that he wrote the true and accurate facts about what he was really thinking because I spent quite a bit of time talking to the General that day and he sure fooled me. I had no idea he was worried about what he should tell the public.

When the press conference, which was the largest and longest the Air Force had held since World War II, convened at 4:00P.M., General Samford made an honest effort to straighten out the Washington National Sightings, but the cards were stacked against him before he started. He had to hedge on many answers to questions from the press because he didn't know the answers. This hedging gave the impression that he was trying to cover up something more than just the fact that his people had fouled up in not fully investigating the sightings. Then he had brought in Captain Roy James from ATIC to handle all the queries about radar. James didn't do any better because he'd just arrived in Washington that morning and didn't know very much more about the sightings than he'd read in the papers. Major Dewey Fournet and Lieutenant Holcomb, who had been at the airport during the sightings, were extremely conspicuous by their absence, especially since it was common knowledge among the press that they weren't convinced the UFO's picked up on radars were weather targets.

But somehow out of this chaotic situation came exactly the result that was intended—the press got off our backs. Captain James's answers about the possibility of the radar targets' being caused by temperature inversions had been construed by the press to mean that this was the Air Force's answer, even though today the twin sightings are still carried as unknowns.

The next morning headlines from Bangor to Bogota read:

AIR FORCE DEBUNKS SAUCERS AS JUST NATURAL PHENOMENA

The Washington National Sightings proved one thing, something that many of us already knew: in order to forestall any more trouble similar to what we'd just been through we always had to get all of the facts and not try to hide them. A great deal of the press's interest was caused by the Air Force's reluctance to give out any information, and the reluctance on the part of the Air Force was caused by simply not having gone out to find the answers.

But had someone gone out and made a more thorough investigation a few big questions would have popped up and taken some of the intrigue out of the two reports. It took me a year to put the question marks together because I just picked up the information as I happened to run across it, but it could have been collected in a day of concentrated effort.

There was some doubt about the visual sighting of the "large fiery- orange-colored sphere" that the tower operators at Andrews AFB saw when the radar operators at National Airport told them they had a target over the Andrews Radio range station. When the tower operators were later interrogated they completely changed their story and said that what they saw was merely a star. They said that on the night of the sighting they "had been excited." (According to astronomical charts, there were no exceptionally bright stars where the UFO was seen over the range station, however. And I heard from a good source that the tower men had been "persuaded" a bit.)

Then the pilot of the F-94C changed his mind even after he'd given the press and later told me his story about vainly trying to intercept unidentified lights. In an official report he says that all he saw was a ground light reflecting off a layer of haze.

Another question mark arose about the lights that the airline pilots saw. Months after the sighting I heard from one of the pilots whom the ARTC controllers called to learn if he could see a UFO. This man's background was also impressive, he had been flying in and out of Washington since 1936. This is what he had to say:

The most outstanding incident happened just after a take-off one night from Washington National. The tower man advised us that there was a UFO ahead of us on the take-off path and asked if we would aid in tracking it down. We were given headings to follow and shortly we were advised that we had passed the UFO and would be given a new heading. None of us in the cockpit had seen anything unusual. Several runs were made; each time the tower man advised us we were passing the UFO we noticed that we were over one certain section of the Potomac River, just east of Alexandria. Finally we were asked to visually check the terrain below for anything which might cause such an illusion. We looked and the only object we could see where the radar had a target turned out to be the Wilson Lines moonlight steamboat trip to Mount Vernon. Whether there was an altitude gimmick on the radar unit at the time I do not know but the radar was sure as hell picking up the steamboat.

The pilot went on to say that there is such a conglomeration of lights around the Washington area that no matter where you look you see a "mysterious light."

Then there was another point: although the radars at Washington National and Andrews overlap, and many of the targets appeared in the overlap area, only once did the three radars simultaneously pick up a target.

The investigation brought out a few more points on the pro side too. We found out that the UFO's frequently visited Washington. On May 23 fifty targets had been tracked from 8:00 p.m. till midnight. They were back on the Wednesday night between the two famous Saturday- night sightings, the following Sunday night, and again the night of the press conference; then during August they were seen eight more times. On several occasions military and civilian pilots saw lights exactly where the radar showed the UFO's to be.

On each night that there was a sighting there was a temperature inversion but it was never strong enough to affect the radar the way inversions normally do. On each occasion I checked the strength of the inversion according to the methods used by the Air Defense Command Weather Forecast Center.

Then there was another interesting fact: hardly a night passed in June, July, and August in 1952 that there wasn't an inversion in Washington, yet the slow-moving, "solid" radar targets appeared on only a few nights.

But the one big factor on the pro side of the question is the people involved—good radar men—men who deal in human lives. Each day they use their radar to bring thousands of people into Washington National Airport and with a responsibility like this they should know a real target from a weather target.

So the Washington National Airport Sightings are still unknowns.

Had the press been aware of some of the other UFO activity in the United States during this period, the Washington sightings might not have been the center of interest. True, they could be classed as good reports but they were not the best that we were getting. In fact, less than six hours after the ladies and gentlemen of the press said "Thank you" to General Samford for his press conference, and before the UFO's could read the newspapers and find out that they were natural phenomena, one of them came down across the Canadian border into Michigan. The incident that occurred that night was one of those that even the most ardent skeptic would have difficulty explaining. I've heard a lot of them try and I've heard them all fail.

At nine-forty on the evening of the twenty-ninth an Air Defense Command radar station in central Michigan started to get plots on a target that was coming straight south across Saginaw Bay on Lake Huron at 625 miles an hour. A quick check of flight plans on file showed that it was an unidentified target.

Three F-94's were in the area just northeast of the radar station, so the ground controller called one of the F-94's and told the pilot to intercept the unidentified target. The F-94 pilot started climbing out of the practice area on an intercept heading that the ground controller gave him. When the F-94 was at 20,000 feet, the ground controller told the pilot to turn to the right and he would be on the target. The pilot started to bring the F-94 around and at that instant both he and the radar operator in the back seat saw that they were turning toward a large bluish-white light, "many times larger than a star." In the next second or two the light "took on a reddish tinge, and slowly began to get smaller, as if it were moving away." Just then the ground controller called and said that he still had both the F-94 and the unidentified target on his scope and that the target had just made a tight 180-degree turn. The turn was too tight for a jet, and at the speed the target was traveling it would have to be a jet if it were an airplane. Now the target was heading back north. The F-94 pilot gave the engine full power and cut in the afterburner to give chase. The radar operator in the back seat got a good radar lock-on. Later he said, "It was just as solid a lock-on as you get from a B-36." The object was at 4 miles range and the F-94 was closing slowly. For thirty seconds they held the lock-on; then, just as the ground controller was telling the pilot that he was closing in, the light became brighter and the object pulled away to break the lock-on. Without breaking his transmission, the ground controller asked if the radar operator still had the lock-on because on the scope the distance between two blips had almost doubled in one sweep of the antenna. This indicated that the unknown target had almost doubled its speed in a matter of seconds.

For ten minutes the ground radar followed the chase. At times the unidentified target would slow down and the F-94 would start to close the gap, but always, just as the F-94 was getting within radar range, the target would put on a sudden burst of speed and pull away from the pursuing jet. The speed of the UFO—for by this time all concerned had decided that was what it was—couldn't be measured too accurately because its bursts of speed were of such short duration; but on several occasions the UFO traveled about 4 miles in one ten- second sweep of the antenna, or about 1,400 miles an hour.

The F-94 was getting low on fuel, and the pilot had to break off the chase a minute or two before the UFO got out of range of the ground radar. The last few plots on the UFO weren't too good but it looked as if the target slowed down to 200 to 300 miles an hour as soon as the F-94 turned around.

What was it? It obviously wasn't a balloon or a meteor. It might have been another airplane except that in 1952 there was nothing flying, except a few experimental airplanes that were far from Michigan, that could so easily outdistance an F-94. Then there was the fact that radar clocked it at 1,400 miles an hour. The F-94 was heading straight for the star Capella, which is low on the horizon and is very brilliant, but what about the radar contacts? Some people said "Weather targets," but the chances of a weather target's making a 180-degree turn just as an airplane turns into it, giving a radar lock-on, then changing speed to stay just out of range of the airplane's radar, and then slowing down when the airplane leaves is as close to nil as you can get.

What was it? A lot of people I knew were absolutely convinced this report was the key—the final proof. Even if all of the thousands of other UFO reports could be discarded on a technicality, this one couldn't be. These people believed that this report in itself was proof enough to officially accept the fact that UFO's were interplanetary spaceships. And when some people refused to believe even this report, the frustration was actually pitiful to see.

As the end of July approached, there was a group of officers in intelligence fighting hard to get the UFO "recognized." At ATIC, Project Blue Book was still trying to be impartial—but sometimes it was difficult.

Chapter Thirteen - Hoax or Horror?

To the military and the public who weren't intimately associated with the higher levels of Air Force Intelligence during the summer of 1952—and few were—General Samford's press conference seemed to indicate the peak in official interest in flying saucers. It did take the pressure off Project Blue Book—reports dropped from fifty per day to ten a day inside of a week—but behind the scenes the press conference was only the signal for an all-out drive to find out more about the UFO. Work on the special cameras continued on a high- priority basis, and General Samford directed us to enlist the aid of top-ranking scientists.

During the past four months we had collected some 750 comparatively well-documented reports, and we hoped that something in these reports might give us a good lead on the UFO. My orders were to tell the scientists to whom we talked that the Air Force was officially still very much interested in the UFO and that their assistance, even if it was only in giving us ideas and comments on the reports, was badly needed. Although the statement of the problem was worded much more loosely, in essence it was, "Do the UFO reports we have collected indicate that the earth is being visited by a people from another planet?"

Such questions had been asked of the scientists before, but not in such a serious vein.

Then a secondary program was to be started, one of "educating" the military. The old idea that UFO reports would die out when the thrill wore off had long been discarded. We all knew that UFO reports would continue to come in and that in order to properly evaluate them we had to have every shred of evidence. The Big Flap had shown us that our chances of getting a definite answer on a sighting was directly proportional to the quality of the information we received from the intelligence officers in the field.

But soon after the press conference we began to get wires from intelligence officers saying they had interpreted the newspaper accounts of General Samford's press conference to mean that we were no longer interested in UFO reports. A few other intelligence officers had evidently also misinterpreted the general's remarks because their reports of excellent sightings were sloppy and incomplete. All of this was bad, so to forestall any misconceived ideas about the future of the Air Force's UFO project, summaries of General Samford's press conference were distributed to intelligence officers. General Samford had outlined the future of the UFO project when he'd said:

"So our present course of action is to continue on this problem with the best of our ability, giving it the attention that we feel it very definitely warrants. We will give it adequate attention, but not frantic attention."

The summary of the press conference straightened things out to some extent and our flow of reports got back to normal.

I was anxious to start enlisting the aid of scientists, as General Samford had directed, but before this could be done we had a backlog of UFO reports that had to be evaluated. During July we had been swamped and had picked off only the best ones. Some of the reports we were working on during August had simple answers, but many were unknowns. There was one report that was of special interest because it was an excellent example of how a UFO report can at first appear to be absolutely unsoluble then suddenly fall apart under thorough investigation. It also points up the fact that our investigation and analysis were thorough and that when we finally stamped a report "Unknown" it was unknown. We weren't infallible but we didn't often let a clue slip by.

At exactly ten forty-five on the morning of August 1, 1952, an ADC radar near Bellefontaine, Ohio, picked up a high-speed unidentified target moving southwest, just north of Dayton. Two F-86's from the 97th Fighter-Interceptor Squadron at Wright-Patterson were scrambled and in a few minutes they were climbing out toward where the radar showed the UFO to be. The radar didn't have any height-finding equipment so all that the ground controller at the radar site could do was to get the two F-86's over or under the target, and then they would have to find it visually.

When the two airplanes reached 30,000 feet, the ground controller called them and told them that they were almost on the target, which was still continuing its southwesterly course at about 525 miles an hour. In a few seconds the ground controller called back and told the lead pilot that the targets of his airplane and the UFO had blended on the radar-scope and that the pilot would have to make a visual search; this was as close in as radar could get him. Then the radar broke down and went off the air.

But at almost that exact second the lead pilot looked up and there in the clear blue sky several thousand feet above him was a silver- colored sphere. The lead pilot pointed it out to his wing man and both of them started to climb. They went to their maximum altitude but they couldn't reach the UFO. After ten minutes of unsuccessful attempts to identify the huge silver sphere or disk—because at times it looked like a disk—one of the pilots hauled the nose of his F-86 up in a stall and exposed several feet of gun camera film. Just as he did this the warning light on his radar gun sight blinked on, indicating that something solid was in front of him—he wasn't photographing a sundog, hallucination, or refracted light.

The two pilots broke off the intercept and started back to Wright- Patterson when they suddenly realized that they were still northwest of the base, in almost the same location they had been when they started the intercept ten minutes before. The UFO had evidently slowed down from the speed that the radar had measured, 525 miles an hour, until it was hovering almost completely motionless.

As soon as the pilots were on the ground, the magazine of film from the gun camera was rushed to the photo lab and developed. The photos showed only a round, indistinct blob—no details—but they were proof that some type of unidentified flying object had been in the air north of Dayton.

Lieutenant Andy Flues was assigned to this one. He checked the locations of balloons and found out that a 20-foot-diameter radiosonde weather balloon from Wright-Patterson had been very near the area when the unsuccessful intercept took place, but the balloon wasn't traveling 525 miles an hour and it couldn't be picked up by the ground radar, so he investigated further. The UFO couldn't have been another airplane because airplanes don't hover in one spot and it was no atmospheric phenomenon. Andy wrote it off as an unknown but it still bothered him; that balloon in the area was mighty suspicious. He talked to the two pilots a half dozen times and spent a day at the radar site at Bellefontaine before he reversed his "Unknown" decision and came up with the answer.

The unidentified target that the radar had tracked across Ohio was a low-flying jet. The jet was unidentified because there was a mix-up and the radar station didn't get its flight plan. Andy checked and found that a jet out of Cleveland had landed at Memphis at about eleven-forty. At ten forty-five this jet would have been north of Dayton on a southwesterly heading. When the ground controller blended the targets of the two F-86's into the unidentified target, they were at 30,000 feet and were looking for the target at their altitude or higher so they missed the low-flying jet—but they did see the balloon. Since the radar went out just as the pilots saw the balloon, the ground controller couldn't see that the unidentified target he'd been watching was continuing on to the southwest. The pilots didn't bother to look around any more once they'd spotted the balloon because they thought they had the target in sight.

The only part of the sighting that still wasn't explained was the radar pickup on the F-86's gun sight. Lieutenant Flues checked around, did a little experimenting, and found out that the small transmitter box on a radiosonde balloon will give an indication on the radar used in F-86 gun sights.

To get a final bit of proof, Lieutenant Flues took the gun camera photos to the photo lab. The two F-86's had been at about 40,000 feet when the photos were taken and the 20-foot balloon was at about 70,000 feet. Andy's question to the photo lab was, "How big should a 20-foot balloon appear on a frame of 16-mm. movie film when the balloon is 30,000 feet away?"

The people in the photo lab made a few calculations and measurements and came up with the answer, "A 20-foot balloon photographed from 30,000 feet away would be the same size as the UFO in the gun camera photos."

By the middle of August, Project Blue Book was back to normal. Lieutenant Flues's Coca-Cola consumption had dropped from twenty bottles a day in mid-July to his normal five. We were all getting a good night's sleep and it was now a rare occasion when my home telephone would ring in the middle of the night to report a new UFO.

But then on the morning of August 20 I was happily taking a shower, getting ready to go to work, when one of these rare occasions occurred and the phone rang—it was the ATIC OD. An operational immediate wire had just come in for Blue Book. He had gone over to the message center and gotten it. He thought that it was important and wanted me to come right out. For some reason he didn't want to read it over the phone, although it was not classified. I decided that if he said so I should come out, so I left in a hurry.

The wire was from the intelligence officer at an air base in Florida. The previous night a scoutmaster and three boy scouts had seen a UFO. The scoutmaster had been burned when he approached too close to the UFO. The wire went on to give a few sketchy details and state that the scoutmaster was a "solid citizen."

I immediately put in a long-distance call to the intelligence officer. He confirmed the data in the wire. He had talked briefly to the scoutmaster on the phone and from all he could gather it was no hoax. The local police had been contacted and they verified the story and the fact of the burns. I asked the intelligence officer to contact the scoutmaster and ask if he would submit to a physical examination immediately. I could imagine the rumors that could start about the scoutmaster's condition, and I wanted proof. The report sounded good, so I told the intelligence officer I'd get down to see him as soon as possible.

I immediately called Colonel Dunn, then chief at ATIC, and gave him a brief rundown. He agreed that I should go down to Florida as soon as possible and offered to try to get an Air Force B-25, which would save time over the airlines.

I told Bob Olsson to borrow a Geiger counter at Wright Field, then check out a camera. I called my wife and asked her to pack a few clothes and bring them out to me. Bob got the equipment, ran home and packed a bag, and in two hours he and I and our two pilots, Captain Bill Hoey and Captain David Douglas, were on our way to Florida to investigate one of the weirdest UFO reports that I came up against.

When we arrived, the intelligence officer arranged for the scoutmaster to come out to the air base. The latter knew we were coming, so he arrived at the base in a few minutes. He was a very pleasant chap, in his early thirties, not at all talkative but apparently willing to co-operate.

While he was giving us a brief personal history, I had the immediate impression that he was telling the truth. He'd lived in Florida all of his life. He'd gone to a private military prep school, had some college, and then had joined the Marines. He told us that he had been in the Pacific most of the war and repeated some rather hairy stories of what he'd been through. After the war he'd worked as an auto mechanic, then gone to Georgia for a while to work in a turpentine plant. After returning to Florida, he opened a gas station, but some hard luck had forced him to sell out. He was now working as a clerk in a hardware store. Some months back a local church had decided to organize a boy scout troop and he had offered to be the scoutmaster.

On the night before the weekly scout meeting had broken up early. He said that he had offered to give four of the boys a ride home. He had let one of the boys out when the conversation turned to a stock car race that was to take place soon. They talked about the condition of the track. It had been raining frequently, and they wondered if the track was flooded, so they drove out to look at it. Then they started south toward a nearby town to take another of the boys home. They took a black-top road about 10 miles inland from the heavily traveled coastal highway that passes through sparsely settled areas of scrub pine and palmetto thickets.

They were riding along when the scoutmaster said that he noticed a light off to his left in the pines. He slowed down and asked the boys if they'd seen it; none of them had. He started to drive on, when he saw the lights again. This time all of the boys saw them too, so he stopped. He said that he wanted to go back into the woods to see what was going on, but that the boys were afraid to stay alone. Again he started to drive on, but in a few seconds decided he had to go back. So he turned the car around, went back, and parked beside the road at a point just opposite where he'd seen the lights.

I stopped him at this point to find out a little bit more about why he'd decided to go back. People normally didn't go running off into palmetto thickets infested with rattlesnakes at night. He had a logical answer. The lights looked like an airplane crashing into the woods some distance away. He didn't believe that was what he saw, but the thought that this could be a possibility bothered him. After all, he had said, he was a scoutmaster, and if somebody was in trouble, his conscience would have bothered him the rest of his life if he hadn't investigated and it had been somebody in need of help.

A fifteen-minute radio program had just started, and he told the boys that he was going to go into the woods, and that if he wasn't back by the time the program ended they should run down the road to a farmhouse that they had passed and get help. He got out and started directly into the woods, wearing a faded denim billed cap and carrying machete and two flashlights. One of the lights was a spare he carried in his back pocket.

He had traveled about 50 yards off the road when he ran into a palmetto thicket, so he stopped and looked for a clear path. But finding none, he started pushing his way through the waist-high tangle of brush.

When he stopped, he recalled later, he had first become aware of an odd odor. He couldn't exactly describe it to us, except to say that it was "sharp" or "pungent." It was very faint, actually more like a subconscious awareness at

158

first. Another sensation he recalled after the incident was a very slight difference in temperature, hardly perceivable, like walking by a brick building in the evening after the sun has set. He hadn't thought anything about either the odor or the heat at the time but later, when they became important, he remembered them.

Paying no attention to these sensations then, he pushed on through the brush, looking up occasionally to check the north star, so that he could keep traveling straight east. After struggling through about 30 yards of palmetto undergrowth, he noticed a change in the shadows ahead of him and stopped to shine the flashlight farther ahead of him to find out if he was walking into a clearing or into one of the many ponds that dot that particular Florida area. It was a clearing.

The boy scouts in the car had been watching the scoutmaster's progress since they could see his light bobbing around. Occasionally he would shine it up at a tree or across the landscape for an instant, so they knew where he was in relation to the trees and thickets. They saw him stop at the edge of the open, shadowed area and shine his light ahead of him.

The scoutmaster then told us that when he stopped this second time he first became consciously aware of the odor and the heat. Both became much more noticeable as he stepped into the clearing. In fact, the heat became almost unbearable or, as he put it, "oppressively moist, making it hard to breathe."

He walked a few more paces and suddenly got a horrible feeling that somebody was watching him. He took another step, stopped, and looked up to find the north star. But he couldn't see the north star, or any stars. Then he suddenly saw that almost the whole sky was blanked out by a large dark shape about 30 feet above him.

He said that he had stood in this position for several seconds, or minutes— he didn't know how long—because now the feeling of being watched had overcome any power of reasoning he had. He managed to step back a few paces, and apparently got out from under the object, because he could see the edge of it silhouetted against the sky.

As he backed up, he said, the air became much cooler and fresher, helping him to think more clearly. He shone his light up at the edge of the object and got a quick but good look. It was circular-shaped and slightly concave on the bottom. The surface was smooth and a grayish color. He pointed to a gray linoleum-topped desk in the intelligence officer's room. "Just like that," he said. The upper part had a dome in the middle, like a turret. The edge of the saucer- shaped object was thick and had vanes spaced about every foot, like buckets on a turbine wheel. Between each vane was a small opening, like a nozzle.

The next reaction that the scoutmaster recalled was one of fury. He wanted to harm or destroy whatever it was that he saw. All he had was a machete, but he wanted to try to jump up and strike at whatever he was looking at. No sooner did he get this idea than he noticed the shadows on the turret change ever so slightly and heard a sound, "like the opening of a well-oiled safe door." He froze where he stood and noticed a small ball of red fire begin to drift toward him. As it floated down it expanded into a cloud of red mist. He dropped his fight and machete, and put his arms over his face. As the mist enveloped him, he passed out.

The boy scouts, in the car, estimated that their scoutmaster had been gone about five minutes when they saw him stop at the edge of the clearing, then walk on in. They saw him stop seconds later, hesitate a few more seconds, then shine the light up in the air. They thought he was just looking at the trees again. The next thing they said they saw was a big red ball of fire engulfing him. They saw him fall, so they spilled out of the car and took off down the road toward the farmhouse.

The farmer and his wife had a little difficulty getting the story out of the boys, they were so excited. All they could get was something about the boys' scoutmaster being in trouble down the road. The farmer called the Florida State Highway Patrol, who relayed the message to the county sheriff's office. In a few minutes a deputy sheriff and the local constable arrived. They picked up the scouts and drove to where their car was parked.

The scoutmaster had no idea of how long he had been unconscious. He vaguely remembered leaning against a tree, the feeling of wet, dew- covered grass, and suddenly regaining his consciousness. His first reaction was to get out to the highway, so he started to run. About halfway through the palmetto thicket he saw a car stop on the highway. He ran toward it and found the deputy and constable with the boys.

He was so excited he could hardly get his story told coherently. Later the deputy said that in all his years as a law-enforcement officer he had never seen anyone as scared as the scoutmaster was as he came up out of the ditch beside the road and walked into the glare of the headlights. As soon as he'd told his story, they all went back into the woods, picking their way around the palmetto thicket. The first thing they noticed was the flashlight, still burning, in a clump of grass. Next to it was a place where the grass was flattened down, as if a person had been lying there. They looked around for the extra light that the scoutmaster had been carrying, but it was gone. Later searches for this missing flashlight were equally fruitless. They marked the spot where the crushed grass was located and left. The constable took the boy scouts home and the scoutmaster followed the deputy to the sheriff's office. On the way to town the scoutmaster said he first noticed that his arms and face burned. When he arrived at the sheriff's office, he found that his arms, face, and cap *were* burned. The deputy called the Air Force.

There were six people listening to his story. Bob Olsson, the two pilots, the intelligence officer, his sergeant, and I. We each had previously agreed to pick one insignificant detail from the story and then re-question the scoutmaster when he had finished. Our theory was that if he had made up the story he would either repeat the details perfectly or not remember what he'd said. I'd used this many times before, and it was a good indicator of a lie. He passed the test with flying colors. His story sounded good to all of us.

We talked for about another hour, discussing the event and his background. He kept asking, "What did I see?"—evidently thinking that I knew. He said that the newspapers were after him, since the sheriff's office had inadvertently leaked the story, but that he had been stalling them off pending our arrival. I told him it was Air Force policy to allow people to say anything they wanted to about a UFO sighting. We had never muzzled anyone; it was his choice. With that, we thanked him, arranged to pick up the cap and machete to take back to Dayton, and sent him home in a staff car.

By this time it was getting late, but I wanted to talk to the flight surgeon who had examined the man that morning. The intelligence officer found him at the hospital and he said he would be right over. His report was very thorough. The only thing he could find out of the ordinary were minor burns on his arms and the back of his hands. There were also indications that the inside of his nostrils might be burned. The degree of burn could be compared to a light sunburn. The hair had also been singed, indicating a flash heat.

The flight surgeon had no idea how this specifically could have happened. It could have even been done with a cigarette lighter, and he took his lighter and singed a small area of his arm to demonstrate. He had been asked only to make a physical check, so that is what he'd done, but he did offer a suggestion. Check his Marine records; something didn't ring true. I didn't quite agree; the story sounded good to me.

The next morning my crew from ATIC, three people from the intelligence office, and the two law officers went out to where the incident had taken place. We found the spot where somebody had apparently been lying and the scoutmaster's path through the thicket. We checked the area with a Geiger counter, as a precautionary measure, not expecting to find anything; we didn't. We went over the area inch by inch, hoping to find a burned match with which a flare or fireworks could have been lighted, drippings from a flare, or anything that shouldn't have been in a deserted area of woods. We looked at the trees; they hadn't been hit by lightning. The blades of grass under which the UFO supposedly hovered were not burned. We found nothing to contradict the story. We took a few photos of the area and went back to town. On the way back we talked to the constable and the deputy. All they could do was to confirm what we'd heard.

We talked to the farmer and his wife, but they couldn't help. The few facts that the boy scouts had given them before they had a chance to talk to their scoutmaster correlated with his story. We talked to the scoutmaster's employer and some of his friends; he was a fine person. We questioned people who might have been in a position to also observe something; they saw nothing. The local citizens had a dozen theories, and we thoroughly checked each one.

He hadn't been struck by lightning. He hadn't run across a still. There was no indication that he'd surprised a gang of illegal turtle butcherers, smugglers, or bootleggers. There was no indication of marsh gas or swamp fire. The mysterious blue lights in the area turned out to be a farmer arc-welding at night. The other flying saucers were the landing lights of airplanes landing at a nearby airport.

To be very honest, we were trying to prove that this was a hoax, but were having absolutely no success. Every new lead we dug up pointed to the same thing, a true story.

We finished our work on a Friday night and planned to leave early Saturday morning. Bob Olsson and I planned to fly back on a commercial airliner, as the B-25 was grounded for maintenance. Just after dinner that night I got a call from the sheriff's office. It was from a deputy I had talked to, not the one who met the scoutmaster coming out of the woods, but another one, who had been very interested in the incident. He had been doing a little independent checking and found that our singed UFO observer's background was not as clean as he led one to believe. He had been booted out of the Marines after a few months for being AWOL and stealing an automobile, and had spent some time in a federal reformatory in Chillicothe, Ohio. The deputy pointed out that this fact alone meant nothing but that he thought I might be interested in it. I agreed.

The next morning, early, I was awakened by a phone call from the intelligence officer. The morning paper carried the UFO story on the front page. It quoted the scoutmaster as saying that "high brass" from Washington had questioned him late into the night. There was no "high brass," just four captains, a second lieutenant, and a sergeant. He knew we were from Dayton because we had discussed who we were and where we were stationed. The newspaper story went on to say that "he, the scoutmaster, and the Air Force knew what he'd seen but he couldn't tell—it would create a national panic." He'd also hired a press agent. I could understand the "high brass from the Pentagon" as literary license by the press, but this "national panic" pitch was too much. I had just about decided to give up on this incident and write it off as "Unknown" until this happened. From all appearances, our scoutmaster was going to make a fast buck on his experience. Just before leaving for Dayton, I called Major Dewey Fournet in the Pentagon and asked him to do some checking.

Monday morning the machete went to the materials lab at Wright-Patterson. The question we asked was, "Is there anything unusual about this machete? Is it magnetized? Is it radioactive? Has it been heated?" No knife was ever tested so thoroughly for so many things. As in using a Geiger counter to check the area over which the UFO had hovered in the Florida woods, our idea was to investigate every possible aspect of the sighting. They found nothing, just a plain, unmagnetized, unradioactive, unheated, common, everyday knife.

The cap was sent to a laboratory in Washington, D.C., along with the scoutmaster's story. Our question here was, "Does the cap in any way (burns, chemicals, etc.) substantiate or refute the story?"

I thought that we'd collected all the items that could be analyzed in a lab until somebody thought of one I'd missed, the most obvious of them all—soil and grass samples from under the spot where the UFO had hovered. We'd had samples, but in the last-minute rush to get back to Dayton they had been left in Florida. I called Florida and they were shipped to Dayton and turned over to an agronomy lab for analysis.

By the end of the week I received a report on our ex-Marine's military and reformatory records. They confirmed a few suspicions and added new facts. They were not complimentary. The discrepancy between what we'd heard about the scoutmaster while we were in Florida and the records was considered a major factor. I decided that we should go back to Florida and try to resolve this discrepancy.

Since it was hurricane season, we had to wait a few days, then sneak back between two hurricanes. We contacted a dozen people in the city where the scoutmaster lived. All of them had known him for some time. We traced him from his early boyhood to the time of the sighting. To be sure that the people we talked to were reliable, we checked on them. The specific things we found out cannot be told since they were given to us in confidence, but we were convinced that the whole incident was a hoax.

We didn't talk to the scoutmaster again but we did talk to all the boy scouts one night at their scout meeting, and they retold how they had seen their scoutmaster knocked down by the ball of fire. The night before, we had gone out to the area of the sighting and, under approximately the same lighting conditions as existed on the night of the sighting, had re-enacted the scene—especially the part where the boy scouts saw their scoutmaster fall, covered with red fire. We found that not even by standing *on top of the car* could you see a person silhouetted in the clearing where the scoutmaster supposedly fell. The rest of their stories fell apart to some extent too. They were not as positive of details as they had been previously.

When we returned to Dayton, the report on the cap had come back. The pattern of the scorch showed that the hat was flat when it was scorched, but the burned holes—the lab found some minute holes we had missed—had very probably been made by an electrical spark. This was all the lab could find.

During our previous visit we repeatedly asked the question, "Was the hat burned before you went into the woods?" and, "Had the cap been ironed?" We had received the same answers each time: "The hat was not burned because we [the boy scouts] were playing with it at the scout meeting and would have noticed the burns," and, "The cap was new; it had not been washed or ironed." It is rumored that the cap was never returned because it was proof of the authenticity of the sighting. The hat wasn't returned simply because the scoutmaster said that he didn't want it back. No secrets, no intrigue; it's as simple as that.

Everyone who was familiar with the incident, except a few people in the Pentagon, were convinced that this was a hoax until the lab called me about the grass samples we'd sent in. "How did the roots get charred?" Roots charred? I didn't even know what my caller was talking about. He explained that when they'd examined the grass they had knocked the dirt and sand off the roots of the grass clumps and found them charred. The blades of grass themselves were not damaged; they had never been heated, except on the extreme tips of the longer blades. These had evidently been bending over touching the ground and were also charred. The lab had duplicated the charring and had found that by placing live grass clumps in a pan of sand and dirt and heating it to about 300 degrees F. over a gas burner the charring could be duplicated. How it was actually done outside the lab they couldn't even guess.

As soon as we got the lab report, we checked a few possibilities ourselves. There were no hot underground springs to heat the earth, no chemicals in the soil, not a thing we found could explain it. The only way it could have been faked would have been to heat the earth from underneath to 300 degrees F., and how do you do this without using big and cumbersome equipment and disturbing the ground? You can't. Only a few people handled the grass specimens: the lab, the intelligence officer in Florida, and I. The lab wouldn't do it as a joke, then write an official report, and I didn't do it. This leaves the intelligence officer; I'm positive that he wouldn't do it. There may be a single answer everyone is overlooking, but as of now the charred grass roots from Florida are still a mystery.

Writing an official report on this incident was difficult. On one side of the ledger was a huge mass of circumstantial evidence very heavily weighted against the scoutmaster's story being true. On our second trip to Florida, Lieutenant Olsson and I heard story after story about the man's aptitude for dreaming up tall tales. One man told us, "If he told me the sun was shining, I'd look up to make sure." There were parts of his story and those of the boy scouts that didn't quite mesh. None of us ever believed the boy scouts were in on the hoax. They were undoubtedly so impressed by the story that they imagined a few things they didn't actually see. The scoutmaster's burns weren't proof of anything; the flight surgeon had duplicated these by burning his own arm with a cigarette lighter. But we didn't make step one in proving the incident to be a hoax. We thought up dozens of ways that the man could have set up the hoax but couldn't prove one.

In the scoutmaster's favor were the two pieces of physical evidence we couldn't explain, the holes burned in the cap and the charred grass roots.

The deputy sheriff who had first told me about the scoutmaster's Marine and prison record had also said, "Maybe this is the one time in his life he's telling the truth, but I doubt it."

So did we; we wrote off the incident as a hoax. The best hoax in UFO history.

Many people have asked why we didn't give the scoutmaster a lie detector test. We seriously considered it and consulted some experts in this field. They advised against it. In some definite types of cases the lie detector will not give valid results. This, they thought, was one of those cases. Had we done it and had he passed on the faulty results, the publicity would have been a headache.

There is one way to explain the charred grass roots, the burned cap, and a few other aspects of the incident. It's pure speculation; I don't believe that it is the answer, yet it is interesting. Since the blades of the grass were not damaged and the ground had not been disturbed, this one way is the only way (nobody has thought of any other way) the soil could have been heated. It could have been done by induction heating.

To quote from a section entitled "Induction Heating" from an electrical engineering textbook:

A rod of solid metal or any electrical conductor, when subjected to an alternating magnetic field, has electromotive forces set up in it. These electromotive forces cause what are known as "eddy currents." A rise in temperature results from "eddy currents."

Induction heating is a common method of melting metals in a foundry.

Replace the "rod of solid metal" mentioned above with damp sand, an electrical conductor, and assume that a something that was generating a powerful alternating magnetic field was hovering over the ground, and you can explain how the grass roots were charred. To get an alternating magnetic field, some type of electrical equipment was needed. Electricity—electrical sparks—the holes burned in the cap "by electric sparks."

UFO propulsion comes into the picture when one remembers Dr. Einstein's unified field theory, concerning the relationship between electro-magnetism and gravitation.

If this alternating magnetic field can heat metal, why didn't everything the scoutmaster had that was metal get hot enough to burn him? He had a flashlight, machete, coins in his pocket, etc. The answer—he wasn't under the UFO for more than a few seconds. He said that when he stopped to really look at it he had backed away from under it. He did feel some heat, possibly radiating from the ground.

To further pursue this line of speculation, the scoutmaster repeatedly mentioned the unusual odor near the UFO. He described it as being "sharp" or "pungent." Ozone gas is "sharp" or "pungent." To quote from a chemistry book, "Ozone is prepared by passing air between two plates which are charged at a high electrical potential." Electrical equipment again. Breathing too high a concentration of ozone gas will also cause you to lose consciousness.

I used to try out this induction heating theory on people to get their reaction. I tried it out one day on a scientist from Rand. He practically leaped at the idea. I laughed when I explained that I thought this theory just *happened* to tie together the unanswered aspects of the incident in Florida and was not the answer; he was slightly perturbed. "What do you want?" he said. "Does a UFO have to come in and land on your desk at ATIC?"

Chapter Fourteen - Digesting the Data

It was soon after we had written a finis to the Case of the Scoutmaster that I went into Washington to give another briefing on the latest UFO developments. Several reports had come in during early August that had been read with a good deal of interest in the military and other governmental agencies. By late August 1952 several groups in Washington were following the UFO situation very closely.

The sighting that had stirred everyone up came from Haneda AFB, now Tokyo International Airport, in Japan. Since the sighting came from outside the U.S., we couldn't go out and investigate it, but the intelligence officers in the Far East Air Force had done a good job, so we had the complete story of this startling account of an encounter with a UFO. Only a few minor questions had been unanswered, and a quick wire to FEAF brought back these missing data. Normally it took up to three months to get routine questions back and forth, but this time the exchange of wires took only a matter of hours.

Several months after the sighting I talked to one of the FEAF intelligence officers who had investigated it, and in his estimation it was one of the best to come out of the Far East.

The first people to see the UFO were two control tower operators who were walking across the ramp at the air base heading toward the tower to start the midnight shift. They were about a half hour early so they weren't in any big hurry to get up into the tower—at least not until they saw a large brilliant light off to the northeast over Tokyo Bay. They stopped to look at the light for a few seconds thinking that it might be an exceptionally brilliant star, but both men had spent many lonely nights in a control tower when they had nothing to look at except stars and they had never seen anything this bright before. Besides, the light was moving. The two men had lined it up with the corner of a hangar and could see that it was continually moving closer and drifting a little off to the right.

In a minute they had run across the ramp, up the several hundred steps to the tower, and were looking at the light through 7x50 binoculars. Both of the men, and the two tower operators whom they were relieving, got a good look at the UFO. The light was circular in shape and had a constant brilliance. It appeared to be the upper portion of a large, round, dark shape which was about four times the diameter of the light itself. As they watched, the UFO moved in closer, or at least it appeared to be getting closer because it became more distinct. When it moved in, the men could see a second and dimmer light on the lower edge of the dark, shadowy portion.

In a few minutes the UFO had moved off to the east, getting dimmer and dimmer as it disappeared. The four tower men kept watching the eastern sky, and suddenly the light began to reappear. It stayed in sight a few seconds, was gone again, and then for the third time it came back, heading toward the air base.

This time one of the tower operators picked up a microphone, called the pilot of a C-54 that was crossing Tokyo Bay, and asked if he could see the light. The pilot didn't see anything unusual.

At 11:45P.M., according to the logbook in the tower, one of the operators called a nearby radar site and asked if they had an unidentified target on their scopes. They did.

The FEAF intelligence officers who investigated the sighting made a special effort to try to find out if the radar's unidentified target and the light were the same object. They deduced that they were since, when the tower operators and the radar operators compared notes over the telephone, the light and the radar target were in the same location and were moving in the same direction.

For about five minutes the radar tracked the UFO as it cut back and forth across the central part of Tokyo Bay, sometimes traveling so slowly that it almost hovered and then speeding up to 300 miles an hour. All of this time the tower operators were watching the light through binoculars. Several times when the UFO approached the radar station—once it came within 10 miles—a radar operator went outside to find out if he could see the light but no one at the radar site ever saw it. Back at the air base the tower operators had called other people and they saw the light. Later on the tower man said that he had the distinct feeling that the light was highly directional, like a spotlight.

Some of the people who were watching thought that the UFO might be a lighted balloon; so, for the sake of comparison, a lighted weather balloon was released. But the light on the balloon was much more "yellowish" than the UFO and in a matter of seconds it had traveled far enough away that the light was no longer visible. This gave the observers a chance to compare the size of the balloon and the size of the dark, shadowy part of the UFO. Had the UFO been 10 miles away it would have been 50 feet in diameter.

Three minutes after midnight an F-94 scrambled from nearby Johnson AFB came into the area. The ground controller sent the F-94 south of Yokohama, up Tokyo Bay, and brought him in "behind" the UFO. The second that the ground controller had the F-94 pilot lined up and told him that he was in line for a radar run, the radar operator in the rear seat of the F-94 called out that he had a lock-on. His target was at 6,000 yards, 10 degrees to the right and 10 degrees below the F-94. The lock-on was held for ninety seconds as the ground controller watched both the UFO and the F-94 make a turn and come toward the ground radar site. Just as the target entered the "ground clutter"—the permanent and solid target near the radar station caused by the radar beam's striking the ground—the lock-on was broken. The target seemed to pull away swiftly from the jet interceptor. At almost this exact instant the tower operators reported that they had lost visual contact with the UFO. The tower called the F-94 and asked if they had seen anything visually during the chase—they hadn't. The F-94 crew stayed in the area ten or fifteen more minutes but couldn't see anything or pick up any more targets on their radar.

Soon after the F-94 left the area, both the ground radar and the tower operators picked up the UFO again. In about two minutes radar called the tower to say that their target had just "broken into three pieces" and that the three "pieces," spaced about a quarter of a mile apart, were leaving the area, going northeast. Seconds later tower operators lost sight of the light.

The FEAF intelligence officers had checked every possible angle but they could offer nothing to account for the sighting.

There were lots of opinions, weather targets for example, but once again the chances of a weather target's being in exactly the same direction as a bright star and having the star appear to move with the false radar target aren't too likely—to say the least. And then the same type of thing had happened twice before inside of a month's time, once in California and once in Michigan.

As one of the men at the briefing I gave said, "It's incredible, and I can't believe it, but those boys in FEAF are in a war—they're veterans—and by damn, I think they know what they're talking about when they say they've never seen anything like this before."

I could go into a long discourse on the possible explanations for this sighting; I heard many, but in the end there would be only one positive answer— the UFO could not be identified as something we knew about. It could have been an interplanetary spaceship. Many people thought this was the answer and were all for sticking their necks out and establishing a category of conclusions for UFO reports and labeling it spacecraft. But the majority ruled, and a UFO remained an *unidentified* flying object.

On my next trip to the Pentagon I spent the whole day talking to Major Dewey Fournet and two of his bosses, Colonel W. A. Adams and Colonel Weldon Smith, about the UFO subject in general. One of the things we talked about was a new approach to the UFO problem—that of trying to prove that the motion of a UFO as it flew through the air was intelligently controlled.

I don't know who would get credit for originating the idea of trying to analyze the motion of the UFO's. It was one of those kinds of ideas that are passed around, with everyone adding a few modifications. We'd been talking about making a study of this idea for a long time, but we hadn't had many reports to work with; but now, with the mass of data that we had accumulated in June and July and August, the prospects of such a study looked promising.

The basic aim of the study would be to learn whether the motion of the reported UFO's was random or ordered. Random motion is an unordered, helter-skelter motion very similar to a swarm of gnats or flies milling around. There is no apparent pattern or purpose to their flight paths. But take, for example, swallows flying around a chimney—they wheel, dart, and dip, but if you watch them closely, they have a definite pattern in their movements—an ordered motion. The definite pattern is intelligently controlled because they are catching bugs or getting in line to go down the chimney.

By the fall of 1952 we had a considerable number of well-documented reports in which the UFO's made a series of maneuvers. If we could prove that these maneuvers were not random, but ordered, it would be proof that the UFO's were things that were intelligently controlled.

During our discussion Major Fournet brought up two reports in which the UFO seemed to know what it was doing and wasn't just aimlessly darting around. One of these was the recent sighting from Haneda AFB, Japan, and the other was the incident that happened on the night of July 29, when an F-94 attempted to intercept a UFO over eastern Michigan. In both cases radar had established the track of the UFO.

In the Haneda Incident, according to the sketch of the UFO's track, each turn the UFO made was constant and the straight "legs" between the turns were about the same length. The sketch of the UFO's flight path as it moved back and forth over Tokyo Bay reminded me very much of the "crisscross" search patterns we used to fly during World War II when we were searching for the crew of a ditched airplane. The only time the UFO seriously deviated from this pattern was when the F- 94 got on its tail.

The Michigan sighting was even better, however. In this case there was a definite reason for every move that the UFO made. It made a 180- degree turn because the F-94 was closing on it head on. It alternately increased and decreased its speed, but every time it did this it was because the F-94 was closing in and it evidently put on speed to pull out ahead far enough to get out of range of the F-94's radar. To say that this motion was random and that it was just a coincidence that the UFO made the 180-degree turn when the F-94 closed in head on and that it was just a coincidence that the UFO speeded up every time the F-94 began to get within radar range is pushing the chance of coincidence pretty hard.

The idea of the motion analysis study sounded interesting to me, but we were so busy on Project Blue Book we didn't have time to do it. So Major Fournet offered to look into it further and I promised him all the help we could give him.

In the meantime my people in Project Blue Book were contacting various scientists in the U.S., and indirectly in Europe, telling them about our data, and collecting opinions. We did this in two ways. In the United States we briefed various scientific meetings and groups. To get the word to the other countries, we enlisted the gratis aid of scientists who were planning to attend conferences or meetings in Europe. We would brief these European-bound scientists on all of the aspects of the UFO problem so they could informally discuss the problem with their European colleagues.

The one thing about these briefings that never failed to amaze me, although it happened time and time again, was the interest in UFO's within scientific circles. As soon as the word spread that Project Blue Book was giving official briefings to groups with the proper security clearances, we had no trouble in getting scientists to swap free advice for a briefing. I might add that we briefed only groups who were engaged in government work and who had the proper security clearances solely because we could discuss any government project that might be of help to us in pinning down the UFO. Our briefings weren't just squeezed in either; in many instances we would arrive at a place to find that a whole day had been set aside to talk about UFO's. And never once did I meet anyone who laughed off the whole subject of flying saucers even though publicly these same people had jovially sloughed off the press with answers of "hallucinations," "absurd," or "a waste of time and money." They weren't wild-eyed fans but they were certainly interested.

Colonel S. H. Kirkland and I once spent a whole day briefing and talking to the Beacon Hill Group, the code name for a collection of some of the world's leading scientists and industrialists. This group, formed to consider and analyze the toughest of military problems, took a very serious interest in our project and gave much good advice. At Los Alamos and again at Sandia Base our briefings were given in auditoriums to standing room only crowds. In addition I gave my briefings at National Advisory Committee for Aeronautics laboratories, at Air Research and Development centers, at Office of Naval Research facilities and at the Air Force University. Then we briefed special groups of scientists.

Normally scientists are a cautious lot and stick close to proven facts, keeping their personal opinions confined to small groups of friends, but when they know that there is a sign on a door that says "Classified Briefing in Progress," inhibitions collapse like the theories that explain all the UFO's away. People say just what they think.

I could jazz up this part of the UFO story as so many other historians of the UFO have and say that Dr. So-and-So believes that the reported flying saucers are from outer space or that Dr. Whositz is firmly convinced that Mars is inhabited. I talked to plenty of Dr. So-and-So's who believed that flying saucers were real and who were absolutely convinced that other planets or bodies in the universe were inhabited, but we were looking for proven facts and not just personal opinions.

However, some of the questions we asked the scientists had to be answered by personal opinions because the exact answers didn't exist. When such questions came up, about all we could do was to try to get the largest and most representative cross section of personal opinions upon which to base our decisions. In this category of questions probably the most frequently discussed was the possibility that other celestial bodies in the universe were populated with intelligent beings. The exact answer to this is that no one knows. But the consensus was that it wouldn't be at all surprising.

All the briefings we were giving added to our work load because UFO reports were still coming in in record amounts. The lack of newspaper publicity after the Washington sightings had had some effect because the number of reports dropped from nearly 500 in July to 175 in August, but this was still far above the normal average of twenty to thirty reports a month.

September 1952 started out with a rush, and for a while it looked as if UFO sightings were on the upswing again. For some reason, we never could determine why, we suddenly began to get reports from all over the southeastern United States. Every morning, for about a week or two, we'd have a half dozen or so new reports. Georgia and Alabama led the field. Many of the reports came from people in the vicinity of the then new super-hush-hush Atomic Energy Commission facility at Savannah River, Georgia. And many were coming from the port city of Mobile, Alabama. Our first thought, when the reports began to pour in, was that the newspapers in these areas were possibly stirring things up with scare stories, but our newspaper clipping service covered the majority of the southern papers, and although we kept looking for publicity, none showed up. In fact, the papers only barely mentioned one or two of the sightings. As they came in, each of the sighting reports went through our identification process; they were checked against all balloon flights, aircraft flights, celestial bodies, and the MO file, but more than half of them came out as unknowns.

When the reports first began to come in, I had called the intelligence officers at all of the major military installations in the Southeast unsuccessfully trying to find out if they could shed any light on the cause of the sightings. One man, the man who was responsible for UFO reports made to Brookley AFB, just outside of Mobile, Alabama, took a dim view of all of the proceedings. "They're all nuts," he said.

About a week later his story changed. It seems that one night, about the fourth night in a row that UFO's had been reported near Mobile, this man and several of his assistants decided to try to see these famous UFO's; about 10:00P.M., the time that the UFO's were usually reported, they were gathered around the telephone in the man's office at Brookley AFB. Soon a report came in. The first question that the investigator who answered the phone asked was, "Can you still see it?"

The answer was "Yes," so the officer took off to see the UFO.

The same thing happened twice more, and two more officers left for different locations. The fourth time the phone rang the call was from the base radar station. They were picking up a UFO on radar, so the boss himself took off. He saw the UFO in air out over Mobile Bay and he saw the return of the UFO on the radarscope.

The next morning he called me at ATIC and for over an hour he told me what had happened. Never have I talked to four more ardent flying saucer believers.

We did quite a bit of work on the combination radar-visual sighting at Brookley. First of all, radar-visual sightings were the best type of UFO sightings we received. There are no explanations for how radar can pick up a UFO target that is being watched visually at the same time. Maybe I should have said there are no proven explanations on how this can happen, because, like everything else associated with the UFO, there was a theory. During the Washington National Sightings several people proposed the idea that the same temperature-inversion layer that was causing the radar beam to bend down and pick up a ground target was causing the target to appear to be in the air. They went on to say that we couldn't get a radar-visual sighting unless the ground target was a truck, car, house, or something else that was lighted and could be seen at a great distance. The second reason the Brookley AFB sighting was so interesting was that it knocked this theory cold.

The radar at Brookley AFB was so located that part of the area that it scanned was over Mobile Bay. It was in this area that the UFO was detected. We thought of the theory that the same inversion layer that bent the radar beam also caused the target to appear to be in the air, and we began to do a little checking. There was a slight inversion but, according to our calculations, it wasn't enough to affect the radar. More important was the fact that in the area where the target appeared there were no targets to pick up—let alone lighted targets. We checked and rechecked and found that at the time of the sighting there were no ships, buoys, or anything else that would give a radar return in the area of Mobile Bay in which we were interested.

Although this sighting wasn't as glamorous as some we had, it was highly significant because it was possible to show that the UFO couldn't have been a lighted surface target.

While we were investigating the sighting we talked to several electronics specialists about our radar-visual sightings. One of the most frequent comments we heard was, "Why do all of these radar- visual sightings occur at night?"

The answer was simple: they don't. On August 1, just before dawn, an ADC radar station outside of Yaak, Montana, on the extreme northern border of the United States, picked up a UFO. The report was very similar to the sighting at Brookley except it happened in the daylight and, instead of seeing a light, the crew at the radar station saw a "dark, cigar-shaped object" right where the radar had the UFO pinpointed.

What these people saw is a mystery to this day.

Late in September I made a trip out to Headquarters, ADC to brief General Chidlaw and his staff on the past few months' UFO activity.

Our plans for periodic briefings, which we had originally set up with ADC, had suffered a bit in the summer because we were all busy elsewhere. They were still giving us the fullest co-operation, but we hadn't been keeping them as thoroughly read in as we would have liked to. I'd finished the briefing and was eating lunch at the officers' club with Major Verne Sadowski, Project Blue Book's liaison officer in ADC Intelligence, and several other officers. I had a hunch that something was bothering these people. Then finally Major Sadowski said, "Look, Rupe, are you giving us the straight story on these UFO's?"

I thought he meant that I was trying to spice things up a little, so I said that since he had copies of most of our reports and had read them, he should know that I was giving them the facts straight across the board.

172

Then one of the other officers at the table cut in, "That's just the point, we do have the reports and we have read them. None of us can understand why Intelligence is so hesitant to accept the fact that something we just don't know about is flying around in our skies— unless you are trying to cover up something big."

Everyone at the table put in his ideas. One radar man said that he'd looked over several dozen radar reports and that his conclusion was that the UFO's couldn't be anything but interplanetary spaceships. He started to give his reasons when another radar man leaped into the conversation.

This man said that he'd read every radar report, too, and that there wasn't one that couldn't be explained as a weather phenomenon—even the radar-visual sightings. In fact, he wasn't even convinced that we had ever gotten such a thing as radar-visual sighting. He wanted to see proof that an object that was seen visually was the same object that the radar had picked up. Did we have it?

I got back into the discussion at this point with the answer. No, we didn't have proof if you want to get technical about the degree of proof needed. But we did have reports where the radar and visual bearings of the UFO coincided almost exactly. Then we had a few reports where airplanes had followed the UFO's and the maneuvers of the UFO that the pilot reported were the same as the maneuvers of the UFO that was being tracked by radar.

A lieutenant colonel who had been sitting quietly by interjected a well-chosen comment. "It seems the difficulty that Project Blue Book faces is what to accept and what not to accept as proof."

The colonel had hit the proverbial nail on its proverbial head.

Then he went on, "Everyone has a different idea of what proof really is. Some people think we should accept a new model of an airplane after only five or ten hours of flight testing. This is enough proof for them that the airplane will fly. But others wouldn't be happy unless it was flight-tested for five or ten years. These people have set an unreasonably high value on the word 'proof.' The answer is somewhere in between these two extremes."

But where is this point when it comes to UFO's?

There was about a thirty-second pause for thought after the colonel's little speech. Then someone asked, "What about these recent sightings at Mainbrace?"

In late September 1952 the NATO naval forces had held maneuvers off the coast of Europe; they were called Operation Mainbrace. Before they had started someone in the Pentagon had half seriously mentioned that Naval Intelligence should keep an eye open for UFO's, but no one really expected the UFO's to show up. Nevertheless, once again the UFO's were their old unpredictable selves—they were there.

On September 20, a U.S. newspaper reporter aboard an aircraft carrier in the North Sea was photographing a carrier take-off in color when he happened to look back down the flight deck and saw a group of pilots and flight deck crew watching something in the sky. He went back to look and there was a silver sphere moving across the sky just behind the fleet of ships. The object appeared to be large, plenty large enough to show up in a photo, so the reporter shot several pictures. They were developed right away and turned out to be excellent. He had gotten the superstructure of the carrier in each one and, judging by the size of the object in each successive photo, one could see that it was moving rapidly.

The intelligence officers aboard the carrier studied the photos. The object looked like a balloon. From its size it was apparent that if it were a balloon, it would have been launched from one of the ships, so the word went out on the TBS radio: "Who launched a balloon?"

The answer came back on the TBS: "Nobody."

Naval Intelligence double-checked, triple-checked and quadruple- checked every ship near the carrier but they could find no one who had launched the UFO.

We kept after the Navy. The pilots and the flight deck crew who saw the UFO had mixed feelings—some were sure that the UFO was a balloon while others were just as sure that it couldn't have been. It was traveling too fast, and although it resembled a balloon in some ways it was far from being iden- tical to the hundreds of balloons that the crew had seen the aerologists launch.

We probably wouldn't have tried so hard to get a definite answer to the Mainbrace photos if it hadn't been for the events that took place during the rest of the operation, I explained to the group of ADC officers.

The day after the photos had been taken six RAF pilots flying a formation of jet fighters over the North Sea saw something coming from the direction of the Mainbrace fleet. It was a shiny, spherical object, and they couldn't recog- nize it as anything "friendly" so they took after it. But in a minute or two they lost it. When they neared their base, one of the pilots looked back and saw that the UFO was now following him. He turned but the UFO also turned, and again it outdistanced the Meteor in a matter of minutes.

Then on the third consecutive day a UFO showed up near the fleet, this time over Topcliffe Aerodrome in England. A pilot in a Meteor was scrambled and managed to get his jet fairly close to the UFO, close enough to see that the object was "round, silvery, and white" and seemed to "rotate around its vertical axis and sort of wobble." But before he could close in to get a really good look it was gone.

It was these sightings, I was told by an RAF exchange intelligence officer in the Pentagon, that caused the RAF to officially recognize the UFO.

By the time I'd finished telling about the Mainbrace Sightings, it was after the lunch hour in the club and we were getting some get-the- hell-out-of-here looks from the waiters, who wanted to clean up the dining room. But before I could suggest that we leave, Major Sadowski repeated his original question— the one that started the whole discussion—"Are you holding out on us?"

I gave him an unqualified "No." We wanted more positive proof, and until we had it, UFO's would remain unidentified flying objects and no more.

The horizontal shaking of heads illustrated some of the group's thinking.

We had plans for getting more positive proof, however, and I said that just as soon as we returned to Major Sadowski's office I'd tell them what we contemplated doing.

We moved out onto the sidewalk in front of the club and, after discussing a few more sightings, went back into the security area to Sadowski's office and I laid out our plans.

First of all, in November or December the U.S. was going to shoot the first H-bomb during Project Ivy. Although this was Top Secret at the time, it was about the most poorly kept secret in history— everybody seemed to know all about it. Some people in the Pentagon had the idea that there were beings, earthly or otherwise, who might be interested in our activities in the Pacific, as they seemed to be in Operation Mainbrace. Consequently Project Blue Book had been directed to get transportation to the test area to set up a reporting net, brief people on how to report, and analyze their reports on the spot.

Secondly, Project Blue Book was working on plans for an extensive system to track UFO's by instruments. Brigadier General Garland, who had been General Samford's Deputy Director for Production and who had been riding herd on the UFO project for General Samford, was now chief at ATIC, having replaced Colonel Dunn, who went to the Air War College. General Garland had long been in favor of trying to get some concrete information, either positive or negative, about the UFO's. This planned tracking system would replace the defraction grid cameras that were still being developed at ATIC.

Thirdly, as soon as we could we were planning to gather together a group of scientists and let them spend a full week or two studying the UFO problem.

When I left ADC, Major Sadowski and crew were satisfied that we weren't just sitting around twiddling our UFO reports.

During the fall of 1952 reports continued to drop off steadily. By December we were down to the normal average of thirty per month, with about 20 per cent of these falling into the "Unknown" category.

Our proposed trip to the Pacific to watch for UFO's during the H- bomb test was canceled at the last minute because we couldn't get space on an airplane. But the crews of Navy and Air Force security forces who did go out to the tests were thoroughly briefed to look for UFO's, and they were given the procedures on how to track and report them. Back at Dayton we stood by to make quick analysis of any reports that might come in—none came. Nothing that fell into the UFO category was seen during the entire Project Ivy series of atomic shots.

By December work on the planning phase of our instrumentation program was completed. During the two months we had been working on it we had considered everything from giving Ground Observer Corps spotters simple wooden tracking devices to building special radars and cameras. We had talked over our problems with the people at Wright Field who knew about missile-tracking equipment, and we had consulted the camera technicians at the Air Force Aerial Reconnaissance Laboratory. Astronomers explained their equipment and the techniques to use, and we went to Rome, New York, and Boston to enlist the aid of the people who develop the Air Force's electronic equipment.

Our final plan called for visual spotting stations to be established all over northern New Mexico. We'd picked this test location because northern New Mexico still consistently produced more reports than any other area in the U.S. These visual spotting stations would be equipped with a sighting device similar to a gun sight on a bomber. All the operator would have to do would be to follow the UFO with the tracking device, and the exact time and the UFO's azimuth and elevation angles would be automatically recorded. The visual spotting stations would all be tied together with an interphone system, so that as soon as the tracker at one station saw something he could alert the other spotters in the area. If two stations tracked the same object, we could immediately compute its speed and altitude.

This visual spotting net would be tied into the existing radar defense net in the Albuquerque-Los Alamos area. At each radar site we proposed that a long focal-length camera be synchronized to the turning radar antenna, so that any time the operator saw a target he could press a button and photograph the portion of the sky exactly where the radar said a UFO was located. These cameras would actually be astronomical telescopes, so that even the smallest light or object could be photographed.

In addition to this photography system we proposed that a number of sets of instruments be set out around the area. Each set would contain instruments to measure nuclear radiation, any disturbances in the earth's magnetic field, and the passage of a body that was giving off heat. The instruments would continually be sending their information to a central "UFO command post," which would also get reports directly from the radars and the visual spotting stations.

This instrumentation plan would cost about $250,000 because we planned to use as much surplus equipment as possible and tie it into existing communications systems, where they already existed. After the setup was established, it would cost about $25,000 a year to operate. At first glance this seemed like a lot of money, but when we figured out how much the UFO project had cost the Air Force in the past and how much it would probably cost in the future, the price didn't seem too bad—especially if we could solve the UFO problem once and for all.

The powers-that-be at ATIC O.K.'d the plan in December and it went to Washington, where it would have to be approved by General Samford before it went to ADC and then back to the Pentagon for higher Air Force official blessing. From all indications it looked as if we would get the necessary blessings.

But the majority of the effort at Project Blue Book during the fall of 1952 had gone toward collecting together all of the bits and pieces of data that we had accumulated over the past year and a half. We had sorted out the best of the "Unknowns" and made studies of certain aspects of the UFO problem, so that when we could assemble a panel of scientists to review the data we could give them the over- all picture, not just a basketful of parts.

Everyone who knew about the proposed panel meeting was eager to get started because everyone was interested in knowing what this panel would have to say. Although the group of scientists wouldn't be empowered to make the final decision, their recommendations were to go to the President if they decided that the UFO's were real. And any recommendations made by the group of names we planned to assemble would carry a lot of weight.

In the Pentagon and at ATIC book was being made on what their recommendations would be. When I put my money down, the odds were 5 to 3 in favor of the UFO.

Chapter Fifteen - The Radiation Story

The idea for gathering together a group of scientists, to whom we referred as our "panel of experts," had been conceived early in 1952— as soon as serious talk about the possibility that the UFO's might be interplanetary spaceships had taken hold in both military and scientific circles. In fact, when Project Grudge was reorganized in the summer of 1951 the idea had been mentioned, and this was the main reason that our charter had said we were to be only a fact-finding group. The people on previous UFO projects had gone off on tangents of speculation about the identity of the UFO's; they first declared that they were spacecraft, then later, in a complete about-face, they took the whole UFO problem as one big belly laugh. Both approaches had gotten the Air Force into trouble. Why they did this I don't know, because from the start we realized that no one at ATIC, in the Air Force, or in the whole military establishment was qualified to give a final yes or no answer to the UFO problem. Giving a final answer would require a serious decision— probably one of the most serious since the beginning of man.

During 1952 many highly qualified engineers and scientists had visited Project Blue Book and had spent a day or two going over our reports. Some were very much impressed with the reports—some had all the answers.

But all of the scientists who read our reports readily admitted that even though they may have thought that the reports did or did not indicate visitors from outer space, they would want to give the subject a good deal more study before they ever committed themselves in writing. Consequently the people's opinions, although they were valuable, didn't give us enough to base a decision upon. We still needed a group to study our material thoroughly and give us written conclusions and recommendations which could be sent to the President if necessary.

Our panel of experts was to consist of six or eight of the top scientists in the United States. We fully realized that even the Air Force didn't have enough "pull" just to ask all of these people to drop the important work they were engaged in and spend a week or two studying our reports. Nor did we want to do it this way; we wanted to be sure that we had something worth while before asking for their valuable time. So, working through other government agencies, we organized a preliminary review panel of four people. All of them were competent scientists and we knew their reputations were such that if they recommended that a certain top scientist sit on a panel to review our material he would do it.

In late November 1952 the preliminary review panel met at ATIC for three days.

When the meeting ended, the group unanimously recommended that a "higher court" be formed to review the case of the UFO. In an hour their recommendation was accepted by higher Air Force authorities, and the men proceeded to recommend the members for our proposed panel. They picked six men who had reputations as being both practical and theoretical scientists and who were known to have no biased opinions regarding the UFO's.

The meeting of the panel, which would be held in Washington, was tentatively scheduled for late December or early January—depending upon when all of the scientists who had been asked to attend would be free. At Project Blue Book activity went into high gear as we made preparations for the meeting. But before we were very far along our preparations were temporarily sidetracked—I got a lead on the facts behind a rumor. Normally we didn't pay attention to rumors, but this one was in a different class.

Ever since the Air Force had become interested in UFO reports, the comment of those who had been requested to look them over and give a professional opinion was that we lacked the type of data "you could get your teeth into." In even our best reports we had to rely upon what someone had seen. I'd been told many times that if we had even one piece of information that was substantiated by some kind of recorded proof—a set of cinetheodolite movies of a UFO, a spectrum photograph, or any other kind of instrumented data that one could sit down and study—we would have no difficulty getting almost any scientist in the world interested in actively helping us find the answer to the UFO riddle.

The rumor that caused me to temporarily halt our preparations for the high-level conference involved data that we might be able to get our teeth into.

This is the way it went.

In the fall of 1949, at some unspecified place in the United States, a group of scientists had set up equipment to measure background radiation, the small amount of harmless radiation that is always present in our atmosphere. This natural radiation varies to a certain degree, but will never increase by any appreciable amount unless there is a good reason.

178

According to the rumor, two of the scientists at the unnamed place were watching the equipment one day when, for no apparent reason, a sudden increase of radiation was indicated. The radiation remained high for a few seconds, then dropped back to normal. The increase over normal was not sufficient to be dangerous, but it definitely was unusual. All indications pointed to equipment malfunction as the most probable explanation. A quick check revealed no obvious trouble with the gear, and the two scientists were about to start a more detailed check when a third member of the radiation crew came rushing into the lab.

Before they could tell the newcomer about the unexplained radiation they had just picked up, he blurted out a story of his own. He had driven to a nearby town, and on his return trip, as he approached the research lab, something in the sky suddenly caught his eye. High in the cloudless blue he saw three silvery objects moving in a V formation. They appeared to be spherical in shape, but he wasn't sure. The first fact that had hit him was that the objects were traveling too fast to be conventional aircraft. He jammed on the brakes, stopped his car, and shut off the engine. No sound. All he could hear was the quiet whir of a generator in the research lab. In a few seconds the objects had disappeared from sight.

After the first two scientists had briefed their excited colleague on the unusual radiation they had detected, the three men asked each other the $64 question: Was there any connection between the two incidents? Had the UFO's caused the excessive radiation?

They checked the time. Knowing almost exactly when the instruments had registered the increased radiation, they checked on how long it took to drive to the lab from the point where the three silver objects had been seen. The times correlated within a minute or two. The three men proceeded to check their radiation equipment thoroughly. Nothing was wrong.

The rumor stopped here. Nothing that I or anyone else on Project Blue Book could find out shed any further light on the source of the story. People associated with projects similar to the research lab that was mentioned in the rumor were sought out and questioned. Many of them had heard the story, but no one could add any new details. The three unknown scientists, at the unnamed lab, in an unknown part of the United States, might as well never have existed. Maybe they hadn't.

Almost a year after I had first heard the UFO-radiation story I got a long-distance call from a friend on the west coast. I had seen him several months before, at which time I told him about this curious rumor and expressed my wish to find out how authentic it was. Now, on the phone, he told me he had just been in contact with two people he knew and they had the whole story. He said they would be in Los Angeles the following night and would like very much to talk to me.

I hated to fly clear to the west coast on what might be a wild-goose chase, but I did. I couldn't afford to run the risk of losing an opportunity to turn that old recurrent rumor into fact.

Twenty hours later I met the two people at the Hollywood Roosevelt Hotel. We talked for several hours that night, and I got the details on the rumor and a lot more that I hadn't bargained for. Both of my informants were physicists working for the Atomic Energy Commission, and were recognized in their fields. They wanted no publicity and I promised them that they would get none. One of the men knew all the details behind the rumor, and did most of the talking. To keep my promise of no publicity, I'll call him the "scientist."

The rumor version of the UFO-radiation story that had been kicking around in Air Force and scientific circles for so long had been correct in detail but it was by no means complete. The scientist said that after the initial sighting had taken place word was spread at the research lab that the next time the instruments registered abnormal amounts of radiation, some of the personnel were to go outside immediately and look for some object in the sky.

About three weeks after the first incident a repetition did occur. While excessive radiation was registering on the instruments in the lab, a lone dark object was seen streaking across the sky. Again the instruments were checked but, as before, no malfunction was found.

After this second sighting, according to the scientist, an investigation was started at the laboratory. The people who made the visual observations weren't sure that the object they had seen couldn't have been an airplane. Someone thought that perhaps some type of radar equipment in the airplane, if that's what the object was, might have affected the radiation-detection equipment. So arrangements were made to fly all types of aircraft over the area with their radar in operation. Nothing unusual happened. All possible types of airborne research equipment were traced during similar flights in the hope that some special equipment not normally carried in aircraft would be found to have caused the jump in radiation. But nothing out of the ordinary occurred during these tests either.

It was tentatively concluded, the scientist continued, that the abnormally high radiation readings were "officially" due to some freakish equipment malfunction and that the objects sighted visually were birds or airplanes. A report to this effect was made to military authorities, but since the conclusion stated that no flying saucers were involved, the report went into some unknown file. Project Blue Book never got it.

Shortly after the second UFO-radiation episode the research group finished its work. It was at this time that the scientist had first become aware of the incidents he related to me. A friend of his, one of the men involved in the sightings, had sent the details in a letter.

As the story of the sightings spread it was widely discussed in scientific circles, with the result that the conclusion, an equipment malfunction, began to be more seriously questioned. Among the scientists who felt that further investigation of such phenomena was in order, were the man to whom I was talking and some of the people who had made the original sightings.

About a year later the scientist and these original investigators were working together. They decided to make a few more tests, on their own time, but with radiation-detection equipment so designed that the possibility of malfunction would be almost nil. They formed a group of people who were interested in the project, and on evenings and weekends assembled and set up their equipment in an abandoned building on a small mountain peak. To insure privacy and to avoid arousing undue interest among people not in on the project, the scientist and his colleagues told everyone that they had formed a mineral club. The "mineral club" deception covered their weekend expeditions because "rock hounds" are notorious for their addiction to scrambling around on mountains in search for specimens.

The equipment that the group had installed in the abandoned building was designed to be self-operating. Geiger tubes were arranged in a pattern so that some idea as to the direction of the radiation source could be obtained. During the original sightings the equipment- malfunction factor could not be definitely established or refuted because certain critical data had not been measured.

To get data on visual sightings, the "mineral club" had to rely on the flying saucer grapevine, which exists at every major scientific laboratory in the country.

By late summer of 1950 they were in business. For the next three months the scientist and his group kept their radiation equipment operating twenty-four hours a day, but the tapes showed nothing except the usual background activity. The saucer grapevine reported sightings in the general area of the tests, but none close to the instrumented mountaintop.

The trip to the instrument shack, which had to be made every two days to change tapes, began to get tiresome for the "rock hounds," and there was some talk of discontinuing the watch.

But persistence paid off. Early in December, about ten o'clock in the morning, the grapevine reported sightings of a silvery, circular- shaped object near the instrument shack. The UFO was seen by several people.

When the "rock hounds" checked the recording tapes in the shack they found that several of the Geiger tubes had been triggered at 10:17A.M. The registered radiation increase was about 100 times greater than the normal background activity.

Three more times during the next two months the "mineral club's" equipment recorded abnormal radiation on occasions when the grapevine reported visual sightings of UFO's. One of the visual sightings was substantiated by radar.

After these incidents the "mineral club" kept its instruments in operation until June 1951, but nothing more was recorded. And, curiously enough, during this period while the radiation level remained normal, the visual sightings in the area dropped off too. The "mineral club" decided to concentrate on determining the significance of the data they had obtained.

Accordingly, the scientist and the group made a detailed study of their mountaintop findings. They had friends working on many research projects throughout the United States and managed to visit and confer with them while on business trips. They investigated the possibility of unusual sunspot activity, but sunspots had been normal during the brief periods of high radiation. To clinch the elimination of sunspots as a cause, their record tapes showed no burst of radiation when sunspot activity had been abnormal.

The "rock hounds" checked every possible research project that might have produced some stray radiation for their instruments to pick up. They found nothing. They checked and rechecked their instruments, but could find no factor that might have induced false readings. They let other scientists in on their findings, hoping that these outsiders might be able to put their fingers on errors that had been overlooked.

Now, more than a year after the occurrence of the mysterious incidents that they had recorded, a year spent in analyzing their data, the "rock hounds" had no answer.

By the best scientific tests that they had been able to apply, the visual sightings and the high radiation had taken place more or less simultaneously.

Intriguing ideas are hard to kill, and this one had more than one life, possibly because of the element of mystery which surrounds the subject of flying saucers. But the scientific mind thrives on taking the mystery out of unexplained events, so it is not surprising that the investigation went on.

According to my friend the scientist, a few people outside the laboratory where the "rock hounds" worked were told about the activities of the "mineral club," and they started radiation- detection groups of their own.

For instance, two graduate astronomy students from a southwestern university started a similar watch, on a modest scale, using a modified standard Geiger counter as their detection unit. They did not build a recorder into their equipment, however, and consequently were forced to man their equipment continuously, which naturally cut down the time they were in operation. On two occasions they reportedly detected a burst of high radiation.

Although the veracity of the two astronomers was not doubted, the scientist felt that the accuracy of their readings was poor because of the rather low quality of their equipment.

The scientist then told me about a far more impressive effort to verify or disprove the findings of the "mineral club." Word of the "rock hounds" and their work had also spread to a large laboratory in the East. An Air Force colonel, on duty at the lab, told the story to some of his friends, and they decided to look personally into the situation.

Fortunately these people were in a wonderful spot to make such an investigation. At their laboratory an extensive survey of the surrounding area was being made. An elaborate system of radiation- detection equipment had been set up for a radius of 100 miles around the lab. In addition, the defenses of the area included a radar net.

Thanks to the flashing of silver eagles, the colonel's group got permission to check the records of the radiation-survey station and to look over the logs of the radar stations. They found instances where, during the same period of time that radiation in the area had been much higher than normal, radar had had a UFO on the scope. These events had occurred during the period from January 1951 until about June 1951.

Upon learning of the tentative but encouraging findings that the colonel's group had dug out of their past records, people on both the radiation-survey crews and at the radar sites became interested in co- operating for further investigation. A tie-in with the local saucer grapevine established a three-way check.

One evening in July, just before sunset, two of the colonel's group were driving home from the laboratory. As they sped along the highway they noticed two cars stopped ahead of them. The occupants were standing beside the road, looking at something in the sky.

The two scientists stopped, got out of their car, and scanned the sky too. Low on the eastern horizon they saw a bright circular object moving slowly north. They watched it for a while, took a few notes, then drove back to the lab.

Some interesting news awaited them there. Radar had picked up an unidentified target near the spot where the scientists in the car had seen the UFO, and it had been traveling north. A fighter had been scrambled, but when it got into the proper area, the radar target was off the scope. The pilot glimpsed something that looked like the reported UFO, but before he could check further he had to turn into the sun to get on an interception course, and he lost the object.

Several days passed before the radiation reports from all stations could be collected. When the reports did come in they showed that stations east of the laboratory, on an approximate line with the radar track, had shown the highest increase in radiation. Stations west of the lab showed nothing.

The possible significance of this well-covered incident spurred the colonel's group to extend and refine their activities. Their idea was to build a radiation-detection instrument in an empty wing tank and hang the tank on an F-47. Then when a UFO was reported they would fly a search pattern in the area and try to establish whether or not a certain sector of the sky was more radioactive than other sectors. Also, they proposed to build a highly directional detector for the F- 47 and attempt actually to track a UFO.

The design of such equipment was started, but many delays occurred. Before the colonel's group could get any of the equipment built, some of the members left the lab for other jobs, and the colonel, who sparked the operation, was himself transferred elsewhere. The entire effort collapsed.

The scientist was not surprised that I hadn't heard the story of the colonel's group. All the people involved, he said, had kept it quiet in order to avoid ridicule. The scientist added that he would be glad to give me all the data he had on the sightings of his "mineral club," and he told me where to get the information about the two astronomers and the colonel's group.

Armed with the scientist's notes and recorder tapes, I left for my office at Wright-Patterson Air Force Base, Dayton.

With the blessings of my chief, I started to run down the rest of the radiation information. The data we had, especially that from the scientist's "mineral club," had been thoroughly analyzed, but we thought that since we now had access to more general data something new and more significant might be found.

First I contacted the government agency for which all of the people involved in these investigations had been working, the scientists who recorded the original incident, the scientist and his "mineral club," the colonel's group, and the rest.

The people in the agency were very co-operative but stressed the fact that the activities I was investigating were strictly the extracurricular affairs of the scientists involved, had no official sanction, and should not be tied in with the agency in any way, shape, or form. This closed-door reaction was typical of how the words "flying saucer" seem to scare some people.

They did help me locate the report on the original incident, however, and since it seemed to be the only existing copy, I arranged to borrow it. About this same time we located the two graduate astronomy students in New Mexico. Both now had their Ph.D.'s and held responsible jobs on highly classified projects. They repeated their story, which I had first heard from the scientist, but had kept no record of their activities.

On one occasion, just before dawn on a Sunday morning, they were on the roof, making some meteorological observations. One of them was listening to the Geiger counter when he detected a definite increase in the clicking.

Just as the frequency of the clicks reached its highest peak—almost a steady buzz—a large fireball, described by them as "spectacular," flashed across the sky. Both of the observers had seen several of the green fireballs and said that this object was similar in all respects except that the color was a brilliant blue-white.

With the disappearance of the fireball, the counter once more settled down to a steady click per second. They added that once before they had detected a similar increase in the frequency of the clicks but had seen nothing in the sky.

In telling their story, both astronomers stressed the point that their data were open to a great deal of criticism, mainly because of the limited instrumentation they had used. We agreed. Still their work tended to support the findings of the more elaborate and systematic radiation investigations.

The gods who watch over the UFO project were smiling about this time, because one morning I got a call from a colonel on Wright- Patterson Air Force Base. He was going to be in our area that morning and planned to stop in to see me.

He arrived in a few minutes and turned out to be none other than the colonel who had headed the group which had investigated UFO's and radiation at the eastern laboratory. He repeated his story. It was the same as I had heard from the scientist, with a few insignificant changes. The colonel had no records of his group's operations, but knew who had them. He promised to get a wire off to the person immediately, which he did.

The answer was a bit disappointing. During the intervening months the data had been scattered out among the members of the colonel's group, and when the group broke up, so did its collection of records.

So all we had to fall back on was the colonel's word, but since he now was heading a top-priority project at Wright, it would be difficult not to believe him.

After obtaining the colonel's story, we collected all available data concerning known incidents in which there seemed to be a correlation between the visual sighting of UFO's and the presence of excess atomic radiation in the area of the sightings.

There was one last thing to do. I wanted to take the dates and times of all the reported radiation increases and check them against all sources of UFO reports. This project would take a lot of leg work and digging, but I felt that it would offer the most positive and complete evidence we could assemble as to whether or not a correlation existed.

Accordingly, we dug into our files, ADC radar logs, press wire service files, newspaper morgues in the sighting area, and the files of individuals who collect data on saucers. Whenever we found a visual report that correlated with a radiation peak we checked it against weather conditions, balloon tracks, astronomical reports, etc.

As soon as the data had all been assembled, I arranged for a group of Air Force consultants to look it over. I got the same old answer— the data still aren't good enough. The men were very much interested in the reports, but when it came time to putting their comments on paper they said, "Not enough conclusive evidence." If in some way the UFO's could have been photographed at the same time that the radiation detectors were going wild, it would have been a different story, they later told me, but with the data I had for them this was the only answer they could give. No one could explain the sudden bursts of radiation, but there was no proof that they were associated with UFO's.

The board's ruling wrote finish to this investigation. I informed the colonel, and he didn't like the decision. Later I passed through the city where the scientist was working. I stopped over a few hours to brief him on the board's decision. He shook his head in disbelief.

It is interesting to note that both the colonel and the scientist reacted in the same way. We're not fools—we were there—we saw it— they didn't. What do they want for proof?

Chapter Sixteen - The Hierarchy Ponders

By early January 1953 the scientists who were to be members of our panel of experts had been contacted and had agreed to sit in judgment of the UFO. In turn, we agreed to give them every detail about the UFO.

We had our best reports for them to read, and we were going to show them the two movies that some intelligence officers considered as the "positive proof"—the Tremonton Movie and the Montana Movie.

When this high court convened on the morning of January 12, the first thing it received was its orders; one of three verdicts would be acceptable:

All UFO reports are explainable as known objects or natural phenomena; therefore the investigation should be permanently discontinued.

The UFO reports do not contain enough data upon which to base a final conclusion. Project Blue Book should be continued in hopes of obtaining better data.

The UFO's are interplanetary spacecraft.

The written verdict, the group was told, would be given to the National Security Council, a council made up of the directors of all U.S. intelligence agencies, and thence it would go to the President of the United States—if they should decide that the UFO's were interplanetary spacecraft.

Because of military regulations, the names of the panel members, like the names of so many other people associated with the UFO story, cannot be revealed. Two of the men had made names for themselves as practical physicists—they could transform the highest theory for practical uses. One of these men had developed the radar that pulled us out of a big hole at the beginning of World War II, and the other had been one of the fathers of the H-bomb. Another of the panel members is now the chief civilian adviser to one of our top military commanders, and another was an astronomer whose unpublished fight to get the UFO recognized is respected throughout scientific circles. There was a man who is noted for his highly theoretical physics and mathematics, and another who had pioneered operations research during World War II. The sixth member of the panel had been honored by the American Rocket Society and the International Astronautical Federation for his work in moving space travel from the Buck Rogers realm to the point of near reality and who is now a rocket expert.

It was an impressive collection of top scientific talent.

During the first two days of the meeting I reviewed our findings for the scientists. Since June 1947, when the first UFO report had been made, ATIC had analyzed 1,593 UFO reports. About 4,400 had actually been received, but all except 1,593 had been immediately rejected for analysis. From our studies, we estimated that ATIC received reports of only 10 per cent of the UFO sightings that were made in the United States, therefore in five and a half years something like 44,000 UFO sightings had been made.

Of the 1,593 reports that had been analyzed by Project Blue Book, and we had studied and evaluated every report in the Air Force files, we had been able to explain a great many. The actual breakdown was like this:

Balloons..18.51%
Known 1.57
Probable 4.99
Possible 11.95
18.51
Aircraft..11.76%
Known 0.98
Probable 7.74
Possible 3.04
11.76
Astronomical Bodies..14.20%
Known 2.79
Probable 4.01
Possible 7.40
14.20
Other 4.21%
Searchlights on clouds, birds, blowing paper, inversions, reflections, etc.
Hoaxes..1.66%
Insufficient data..22.72%
(In addition to those initially eliminated)
Unknowns..26.94%

By using the terms "Known," "Probable," and "Possible," we were able to differentiate how positive we were of our conclusions. But even in the "Possible" cases we were, in our own minds, sure that we had identified the reported UFO.

And who made these reports? Pilots and air crews made 17.1 per cent from the air. Scientists and engineers made 5.7 per cent, airport control tower operators made an even 1.0 per cent of the reports, and 12.5 per cent of the total were radar reports. The remaining 63.7 per cent were made by military and civilian observers in general.

The reports that we were interested in were the 26.94 per cent or 429 "Unknowns," so we had studied them in great detail. We studied the reported colors of the UFO's, the shapes, the directions they were traveling, the times of day they were observed, and many more details, but we could find no significant pattern or trends. We did find that the most often reported shape was elliptical and that the most often reported color was white or "metallic." About the same number of UFO's were reported as being seen in daytime as at night, and the direction of travel equally covered the sixteen cardinal headings of the compass.

Seventy per cent of the "Unknowns" had been seen visually from the air; 12 per cent had been seen visually from the ground; 10 per cent had been picked up by ground or airborne radar; and 8 per cent were combination visual-radar sightings.

In the over-all total of 1,593 sightings women made two reports for every one made by a man, but in the "Unknowns" the men beat out women ten to one.

There were two other factors we could never resolve, the frequency of the sightings and their geographical distribution. Since the first flurry of reports in July of 1947, each July brought a definite peak in reports; then a definite secondary peak occurred just before each Christmas. We plotted these peaks in sightings against high tides, world-wide atomic tests, the positions of the moon and planets, the general cloudiness over the United States, and a dozen and one other things, but we could never say what caused more people to see UFO's at certain times of the year.

Then the UFO's were habitually reported from areas around "technically interesting" places like our atomic energy installations, harbors, and critical manufacturing areas. Our studies showed that such vital military areas as Strategic Air Command and Air Defense Command bases, some A-bomb storage areas, and large military depots actually produced fewer reports than could be expected from a given area in the United States. Large population centers devoid of any major "technically interesting" facilities also produced few reports.

According to the laws of normal distribution, if UFO's are not intelligently controlled vehicles, the distribution of reports should have been similar to the distribution of population in the United States—it wasn't.

Our study of the geographical locations of sightings also covered other countries. The U.S. by no means had a curb on the UFO market.

In all of our "Unknown" reports we never found one measurement of size, speed, or altitude that could be considered to be even fairly accurate. We could say only that some of the UFO's had been traveling pretty fast.

As far as radar was concerned, we had reports of fantastic speeds— up to 50,000 miles an hour—but in all of these instances there was some doubt as to exactly what caused the target. The highest speeds reported for our combination radar-visual sightings, which we considered to be the best type of sighting in our files, were 700 to 800 miles an hour.

We had never picked up any "hardware"—any whole saucers, pieces, or parts—that couldn't be readily identified as being something very earthly. We had a contract with a materials-testing laboratory, and they would analyze any piece of material that we found or was sent to us. The tar-covered marble, aluminum broom handle, cow manure, slag, pieces of plastic balloon, and the what-have-you that we did receive and analyze only served to give the people in our material lab some practice and added nothing but laughs to the UFO project.

The same went for the reports of "contacts" with spacemen. Since 1952 a dozen or so people have claimed that they have talked to or ridden with the crews of flying saucers. They offer affidavits, pieces of material, photographs, and other bits and pieces of junk as proof. We investigated some of these reports and could find absolutely no fact behind the stories.

We had a hundred or so photos of flying saucers, both stills and movies. Many were fakes—some so expert that it took careful study by photo interpreters to show how the photos had been faked. Some were the crudest of fakes, automobile hub caps thrown into the air, homemade saucers suspended by threads, and just plain retouched negatives. The rest of the still photos had been sent in by well- meaning citizens who couldn't recognize a light flare of flaw in the negative, or who had chanced to get an excellent photo of a sundog or mirage.

But the movies that were sent in to us were different. In the first place, it takes an expert with elaborate equipment to fake a movie. We had or knew about four strips of movie film that fell into the "Unknown" category. Two were the cinetheodolite movies that had been taken at White Sands Proving Ground in April and May of 1950, one was the Montana Movie and the last was the Tremonton Movie. These latter two had been subjected to thousands of hours of analysis, and since we planned to give the panel of scientists more thorough reports on them on Friday, I skipped over their details and went to the next point I wanted to cover—theories.

Periodically throughout the history of the UFO people have come up with widely publicized theories to explain all UFO reports. The one that received the most publicity was the one offered by Dr. Donald Menzel of Harvard University. Dr. Menzel, writing in *Time, Look,* and later in his *Flying Saucers,* claimed that all UFO reports could be explained as various types of light phenomena. We studied this theory thoroughly because it did seem to have merit. Project Bear's physicists studied it. ATIC's scientific consultants studied it and discussed it with several leading European physicists whose specialty was atmospheric physics. In general the comments that Project Blue Book received were, "He's given the subject some thought but his explanations are not the panacea."

And there were other widely publicized theories. One man said that they were all skyhook balloons, but we knew the flight path of every skyhook balloon and they were seldom reported as UFO's. Their little brothers, the weather balloons, caused us a great deal more trouble.

The Army Engineers took a crack at solving the UFO problem by making an announcement that a scientist in one of their laboratories had duplicated a flying saucer in his laboratory. Major Dewey Fournet checked into this one. It had all started out as a joke, but it was picked up as fact and the scientist was stuck with it. He gained some publicity but lost prestige because other scientists wondered just how competent the man really was to try to pass off such an answer.

All in all, the unsolicited assistance of theorists didn't help us a bit, I told the panel members. Some of them were evidently familiar with the theories because they nodded their heads in agreement.

The next topic I covered in my briefing was a question that came up quite frequently in discussions of the UFO: Did UFO reports actually start in 1947? We had spent a great deal of time trying to resolve this question. Old newspaper files, journals, and books that we found in the Library of Congress contained many reports of odd things being seen in the sky as far back as the Biblical times. The old Negro spiritual says, "Ezekiel saw a wheel 'way up in the middle of the air." We couldn't substantiate Ezekiel's sighting because many of the very old reports of odd things observed in the sky could be explained as natural phenomena that weren't fully understood in those days.

The first documented reports of sightings similar to the UFO sightings as we know them today appeared in the newspapers of 1896. In fact, the series of sightings that occurred in that year and the next had many points of similarity with the reports of today.

The sightings started in the San Francisco Bay area on the evening of November 22, 1896, when hundreds of people going home from work saw a large, dark, "cigar-shaped object with stubby wings" traveling northwest across Oakland.

Within hours after the mystery craft had disappeared over what is now the northern end of the Golden Gate Bridge, the stories of people in other northern California towns began to come in on the telegraph wires. The citizens of Santa Rosa, Sacramento, Chico, and Red Bluff— several thousand of them— saw it.

I tried to find out if the people in these outlying communities saw the UFO before they heard the news from the San Francisco area or afterward, but trying to run down the details of a fifty-six-year-old UFO report is almost hopeless. Once while I was on a trip to Hamilton AFB I called the offices of the San Francisco *Chronicle* and they put me in touch with a retired employee who had worked on a San Francisco paper in 1896. I called the old gentleman on the phone and talked to him for a long time. He had been a copy boy at the time and remembered the incident, but time had canceled out the details. He did tell me that he, the editor of the paper, and the news staff had seen "the ship," as he referred to the UFO. His story, even though it was fifty-six years old, smacked of others I'd heard when he said that no one at the newspaper ever told anyone what they had seen; they didn't want people to think that they were "crazy."

On November 30 the mystery ship was back over the San Francisco area and those people who had maintained that people were being fooled by a wag in a balloon became believers when the object was seen moving into the wind.

For four months reports came in from villages, cities, and farms in the West; then the Midwest, as the airship "moved eastward." In early April of 1897 people in Iowa, Nebraska, Missouri, Wisconsin, Minnesota, and Illinois reported seeing it. On April 10 it was reported to be over Chicago. Reports continued to come in to the newspapers until about April 20; then it, or stories about it, were gone. Literally thousands of people had seen it before the last report clicked in over the telegraph wires.

A study of the hundreds of newspaper accounts of this sighting that rocked the world in the late 1890's was interesting because the same controversies that arose then exist now. Those who hadn't seen the stubby-winged, cigar-shaped "craft" said, "Phooey," or the nineteenth- century version thereof. Those who had seen it were almost ready to do battle to uphold their integrity. Some astronomers loudly yelled, "Venus," "Jupiter," and "Alpha Orionis" while others said, "We saw it." Thomas Edison, *the* man of science of the day, disclaimed any knowledge of the mystery craft. "I prefer to devote my time to objects of commercial value," he told a New York *Herald* reporter. "At best airships would only be toys."

Thomas—you goofed on that prediction.

I had one more important point to cover before I finished my briefing and opened the meeting to a general question-and-answer session.

During the past year and a half we had had several astronomers visit Project Blue Book, and they were not at all hesitant to give us their opinions but they didn't care to say much about what their colleagues were thinking, although they did indicate that they were thinking. We decided that the opinions and comments of astronomers would be of value, so late in 1952 we took a poll. We asked an astronomer, whom we knew to be unbiased about the UFO problem and who knew every outstanding astronomer in the United States, to take a trip and talk to his friends. We asked him not to make a point of asking about the UFO but just to work the subject into a friendly conversation. This way we hoped to get a completely frank opinion. To protect his fellow astronomers, our astronomer gave them all code names and he kept the key to the code.

The report we received expressed the detailed opinions of forty-five recognized authorities. Their opinions varied from that of Dr. C, who regarded the UFO project as a "silly waste of money to investigate an even sillier subject," to Dr. L, who has spent a great deal of his own valuable time personally investigating UFO reports because he believes that they are something "real." Of the forty-five astronomers who were interviewed, 36 per cent were not at all interested in the UFO reports, 41 per cent were interested to the point of offering their services if they were ever needed, and 23 per cent thought that the UFO's were a much more serious problem than most people recognized.

None of the astronomers, even during a friendly discussion, admitted that he thought the UFO's could be interplanetary vehicles. All of those who were interested would only go so far as to say, "We don't know what they are, but they're something real."

During the past few years I have heard it said that if the UFO's were really "solid objects" our astronomers would have seen them. Our study shed some light on this point—astronomers have seen UFO's. None of them has ever seen or photographed anything resembling a UFO through his telescope, but 11 per cent of the forty-five men had seen something that they couldn't explain. Although, technically speaking, these sightings were no better than hundreds of others in our files as far as details were concerned, they were good because of the caliber of the observer. Astronomers know what is in the sky.

It is interesting to note that out of the representative cross section of astronomers, five of them, or 11 per cent, had sighted UFO's. For a given group of people this is well above average. To check this point, the astronomer who was making our study picked ninety people at random—people he met while traveling—and got them into a conversation about flying saucers. These people were his "control group," to borrow a term from the psychologists. Although the percentage of people who were interested in UFO's was higher for the control group than for the group of astronomers, only 41 per cent of the astronomers were interested while 86 per cent of the control group were interested; 11 per cent of the astronomers had seen UFO's, while only about 1 per cent of the control group had seen one. This seemed to indicate that as a group astronomers see many more UFO's than the average citizen.

When I finished my briefing, it was too late to start the question- and-answer session, so the first day's meeting adjourned. But promptly at nine o'clock the next morning the group was again gathered, and from the looks of the list of questions some of them had, they must have been thinking about UFO's all night.

One of the first questions was about the results of photography taken by the pairs of huge "meteorite patrol" cameras that are located in several places throughout North America. Did they ever photograph a UFO? The cameras, which are in operation almost every clear night, can photograph very dim lights, and once a light is photographed its speed and altitude can be very accurately established. If there were any objects giving off light as they flew through our atmosphere, there is a chance that these cameras might have photographed them. But they hadn't.

At first this seemed to be an important piece of evidence and we had just about racked this fact up as a definite score against the UFO when we did a little checking. If the UFO had been flying at an altitude of 100 miles, the chances of its being picked up by the cameras would be good, but the chances of photographing something flying any lower would be less.

This may account for the fact that while our "inquiring astronomer" was at the meteorite patrol camera sites, he talked to an astronomer who had seen a UFO while operating one of the patrol cameras.

Many people have asked why our astronomers haven't seen anything through their big telescopes. They are focused light-years away and their field of vision is so narrow that even if UFO's did exist and littered the atmosphere they wouldn't be seen.

Another question the panel had was about Orson Welles' famous *War of the Worlds* broadcast of October 1938, which caused thousands of people to panic. Had we studied this to see if there were any similarities between it and the current UFO reporting?

We had.

Our psychologist looked into the matter and gave us an opinion—to make a complete study and get a positive answer would require an effort that would dwarf the entire UFO project. But he did have a few comments. There were many documented cases in which a series of innocent circumstances triggered by the broadcast had caused people to completely lose all sense of good judgment—to panic. There were some similar reports in our UFO files.

But we had many reports in which people reported UFO's and obviously hadn't panicked. Reports from pilots who had seen mysterious lights at night and, thinking that they might be a cockpit reflection, had turned off all their cockpit lights. Or the pilots who turned and rolled their airplanes to see if they could change the angle of reflection and get rid of the UFO. Or those pilots who climbed and dove thousands of feet and then leveled out to see if the UFO would change its relative position to the airplane. Or the amateur astronomer who made an excellent sighting and before he reluctantly reported it as a UFO had talked to a half dozen professional astronomers and physicists in hopes of finding an explanation. All of these people were thinking clearly, questioning themselves as to what the sightings could be; then trying to answer their questions. These people weren't panicked.

The question-and-answer period went on for a full day as the scientists dug into the details of the general facts I had given them in my briefing.

The following day and a half was devoted to reviewing and discussing fifty of our best sighting reports that we had classed as "Unknowns."

The next item on the agenda, when the panel had finished absorbing all of the details of the fifty selected top reports, was a review of a very hot and very highly controversial study. It was based on the idea that Major Dewey Fournet and I had talked about several months before—an analysis of the motions of the reported UFO's in an attempt to determine whether they were intelligently controlled. The study was hot because it wasn't official and the reason it wasn't official was because it was so hot. It concluded that the UFO's were interplanetary spaceships. The report had circulated around high command levels of intelligence and it had been read with a good deal of interest. But even though some officers at command levels just a notch below General Samford bought it, the space behind the words "Approved by" was blank—no one would stick his neck out and officially send it to the top.

Dewey Fournet, who had completed his tour of active duty in the Air Force and was now a civilian, was called from Houston, Texas, to tell the scientists about the study since he had worked very closely with the group that had prepared it.

The study covered several hundred of our most detailed UFO reports. By a very critical process of elimination, based on the motion of the reported UFO's, Fournet told the panel how he and any previous analysis by Project Blue Book had been disregarded and how those reports that could have been caused by any one of the many dozen known objects—balloons, airplanes, astronomical bodies, etc., were sifted out. This sifting took quite a toll, and the study ended up with only ten or twenty reports that fell into the "Unknown" category. Since such critical methods of evaluation had been used, these few reports proved beyond a doubt that the UFO's were intelligently controlled by persons with brains equal to or far surpassing ours.

The next step in the study, Fournet explained, was to find out where they came from. "Earthlings" were eliminated, leaving the final answer—spacemen.

Both Dewey and I had been somewhat worried about how the panel would react to a study with such definite conclusions. But when he finished his presentation, it was obvious from the tone of the questioning that the men were giving the conclusions serious thought. Fournet's excellent reputation was well known.

On Friday morning we presented the feature attractions of the session, the Tremonton Movie and the Montana Movie. These two bits of evidence represented the best photos of UFO's that Project Blue Book had to offer. The scientists knew about them, especially the Tremonton Movie, because since late July they had been the subject of many closed-door conferences. Generals, admirals, and GS-16's had seen them at "command performances," and they had been flown to Kelly AFB in Texas to be shown to a conference of intelligence officers from all over the world. Two of the country's best military photo laboratories, the Air Force lab at Wright Field and the Navy's lab at Anacostia, Maryland, spent many hours trying to prove that the UFO's were balloons, airplanes, or stray light reflections, but they failed—the UFO's were true unknowns. The possibility that the movie had been faked was considered but quickly rejected because only a Hollywood studio with elaborate equipment could do such a job and the people who filmed the movies didn't have this kind of equipment.

The Montana Movie had been taken on August 15, 1950, by Nick Mariana, the manager of the Great Falls baseball team. It showed two large bright lights flying across the blue sky in an echelon formation. There were no clouds in the movie to give an indication of the UFO's speed, but at one time they passed behind a water tower. The lights didn't show any detail; they appeared to be large circular objects.

Mariana had sent his movies to the Air Force back in 1950, but in 1950 there was no interest in the UFO so, after a quick viewing, Project Grudge had written them off as "the reflections from two F-94 jet fighters that were in the area."

In 1952, at the request of the Pentagon, I reopened the investigation of the Montana Movie. Working through an intelligence officer at the Great Falls AFB, I had Mariana reinterrogated and obtained a copy of his movie, which I sent to the photo lab.

When the photo lab got the movie, they had a little something to work with because the two UFO's had passed behind a reference point, the water tower.

Their calculations quickly confirmed that the objects were not birds, balloons, or meteors. Balloons drift with the wind and the wind was not blowing in the direction that the two UFO's were traveling. No exact speeds could be measured, but the lab could determine that the lights were traveling too fast to be birds and too slow to be meteors.

This left airplanes as the only answer. The intelligence officer at Great Falls had dug through huge stacks of files and found that only two airplanes, two F-94's, were near the city during the sighting and that they had landed about two minutes afterwards. Both Mariana and his secretary, who had also seen the UFO's, had said that the two jets had appeared in another part of the sky only a minute or two after the two UFO's had disappeared in the southeast. This in itself would eliminate the jets as candidates for the UFO's, but we wanted to double-check. The two circular lights didn't look like F-94's, but anyone who has done any flying can tell you that an airplane so far away that it can't be seen can suddenly catch the sun's rays and make a brilliant flash.

First we studied the flight paths of the two F-94's. We knew the landing pattern that was being used on the day of the sighting, and we knew when the two F-94's landed. The two jets just weren't anywhere close to where the two UFO's had been. Next we studied each individual light and both appeared to be too steady to be reflections.

We drew a blank on the Montana Movie—it was an unknown.

We also drew a blank on the Tremonton Movie, a movie that had been taken by a Navy Chief Photographer, Warrant Officer Delbert C. Newhouse, on July 2, 1952.

Our report on the incident showed that Newhouse, his wife, and their two children were driving to Oakland, California, from the east coast on this eventful day. They had just passed through Tremonton, Utah, a town north of Salt Lake City, and had traveled about 7 miles on U.S. Highway 30S when Mrs. Newhouse noticed a group of objects in the sky. She pointed them out to her husband; he looked, pulled over to the side of the road, stopped the car, and jumped out to get a better look. He didn't have to look very long to realize that something highly unusual was taking place because in his twenty-one years in the Navy and 2,000 hours' flying time as an aerial photographer, he'd never seen anything like this. About a dozen shiny disklike objects were "milling around the sky in a rough formation."

Newhouse had his movie camera so he turned the turret around to a 3-inch telephoto lens and started to photograph the UFO's. He held the camera still and took several feet of film, getting all of the bright objects in one photo. All of the UFO's had stayed in a compact group from the time the Newhouse family had first seen them, but just before they disappeared over the western horizon one of them left the main group and headed east. Newhouse swung his camera around and took several shots of it, holding his camera steady and letting the UFO pass through the field of view before it disappeared in the east.

When I received the Tremonton films I took them right over to the Wright Field photo lab, along with the Montana Movie, and the photo technicians and I ran them twenty or thirty times. The two movies were similar in that in both of them the objects appeared to be large circular lights—in neither one could you see any detail. But, unlike the Montana Movie, the lights in the Tremonton Movie would fade out, then come back in again. This fading immediately suggested airplanes reflecting light, but the roar of a king-sized dogfight could have been heard for miles and the Newhouse family had heard no sound. We called in several fighter pilots and they watched the UFO's circling and darting in and out in the cloudless blue sky. Their unqualified comment was that no airplane could do what the UFO's were doing.

Balloons came under suspicion, but the lab eliminated them just as quickly by studying the kind of a reflection given off by a balloon— it is a steady reflection since a balloon is spherical. Then, to further scuttle the balloon theory, clusters of balloons are tied together and don't mill around. Of course, the lone UFO that took off to the east by itself was the biggest argument against balloons.

Newhouse told an intelligence officer from the Western Air Defense Forces that he had held his camera still and let this single UFO fly through the field of view, so the people in the lab measured its angular velocity. Unfortunately there were no clouds in the sky, nor was he able to include any of the ground in the pictures, so our estimates of angular velocity had to be made assuming that the photographer held his camera still. Had the lone UFO been 10 miles away it would have been traveling several thousand miles an hour.

After studying the movies for several weeks, the Air Force photo lab at Wright Field gave up. All they had to say was, "We don't know what they are but they aren't airplanes or balloons, and we don't think they are birds."

While the lab had been working on the movies at Wright Field, Major Fournet had been talking to the Navy photo people at Anacostia; they thought they had some good ideas on how to analyze the movies, so as soon as we were through with them I sent them to Major Fournet and he took them over to the Navy lab.

The Navy lab spent about two months studying the films and had just completed their analysis. The men who had done the work were on hand to brief the panel of scientists on their analysis after the panel had seen the movies.

We darkened the room and I would imagine that we ran each film ten times before every panel member was satisfied that he had seen and could remember all of the details. We ran both films together so that the men could compare them.

The Navy analysts didn't use the words "interplanetary spacecraft" when they told of their conclusions, but they did say that the UFO's were intelligently controlled vehicles and that they weren't airplanes or birds. They had arrived at this conclusion by making a frame-by-frame study of the motion of the lights and the changes in the lights' intensity.

When the Navy people had finished with their presentation, the scientists had questions. None of the panel members were trying to find fault with the work the Navy people had done, but they weren't going to accept the study until they had meticulously searched for every loophole. Then they found one.

In measuring the brilliance of the lights, the photo analysts had used an instrument called a densitometer. The astronomer on the panel knew all about measuring the density of an extremely small photographic image with a densitometer because he did it all the time in his studies of the stars. And the astronomer didn't think that the Navy analysts had used the correct technique in making their measurements. This didn't necessarily mean that their data were all wrong, but it did mean that they should recheck their work.

When the discussion of the Navy's report ended, one of the scientists asked to see the Tremonton Movie again; so I had the projectionists run it several more times. The man said that he thought the UFO's could be sea gulls soaring on a thermal current. He lived in Berkeley and said that he'd seen gulls high in the air over San Francisco Bay. We had thought of this possibility several months before because the area around the Great Salt Lake is inhabited by large white gulls. But the speed of the lone UFO as it left the main group had eliminated the gulls. I pointed this out to the physicist. His answer was that the Navy warrant officer might have thought he had held the camera steady, but he could have "panned with the action" unconsciously. This would throw all of our computations 'way off. I agreed with this, but I couldn't agree that they were sea gulls.

But several months later I was in San Francisco waiting for an airliner to Los Angeles and I watched gulls soaring in a cloudless sky. They were "riding a thermal," and they were so high that you couldn't see them until they banked just a certain way; then they appeared to be a bright white flash, much larger than one would expect from sea gulls. There was a strong resemblance to the UFO's in the Tremonton Movie. But I'm not sure that this is the answer.

The presentation of the two movies ended Project Blue Book's part of the meeting. In five days we had given the panel of scientists every pertinent detail in the history of the UFO, and it was up to them to tell us if they were real—some type of vehicle flying through our atmosphere. If they were real, then they would have to be spacecraft because no one at the meeting gave a second thought to the possibility that the UFO's might be a supersecret U.S. aircraft or a Soviet development. The scientists knew everything that was going on in the U.S. and they knew that no country in the world had developed their technology far enough to build a craft that would perform as the UFO's were reported to do. In addition, we were spending billions of dollars on the research and development and the procurement of airplanes that were just nudging the speed of sound. It would be absurd to think that these billions were being spent to cover the existence of a UFO-type weapon. And it would be equally absurd to think that the British, French, Russians or any other country could be far enough ahead of us to have a UFO.

The scientists spent the next two days pondering a conclusion. They reread reports and looked at the two movies again and again, they called other scientists to double-check certain ideas that they had, and they discussed the problem among themselves. Then they wrote out their conclusions and each man signed the document. The first paragraph said:

We as a group do not believe that it is impossible for some other celestial body to be inhabited by intelligent creatures. Nor is it impossible that these creatures could have reached such a state of development that they could visit the earth. However, there is nothing in all of the so-called "flying saucer" reports that we have read that would indicate that this is taking place.

The Tremonton Movie had been rejected as proof but the panel did leave the door open a crack when they suggested that the Navy photo lab redo their study. But the Navy lab never rechecked their report, and it was over a year later before new data came to light.

After I got out of the Air Force I met Newhouse and talked to him for two hours. I've talked to many people who have reported UFO's, but few impressed me as much as Newhouse. I learned that when he and his family first saw the UFO's they were close to the car, much closer than when he took the movie. To use Newhouse's own words, "If they had been the size of a B-29 they would have been at 10,000 feet altitude." And the Navy man and his family had taken a good look at the objects—they looked like "two pie pans, one inverted on the top of the other!" He didn't just *think* the UFO's were disk-shaped; he *knew* that they were; he had plainly seen them. I asked him why he hadn't told this to the intelligence officer who interrogated him. He said that he had. Then I remembered that I'd sent the intelligence officer a list of questions I wanted Newhouse to answer. The question "What did the UFO's look like?" wasn't one of them because when you have a picture of something you don't normally ask what it looks like. Why the intelligence officer didn't pass this information on to us I'll never know.

The Montana Movie was rejected by the panel as positive proof because even though the two observers said that the jets were in another part of the sky when they saw the UFO's and our study backed them up, there was still a chance that the two UFO's could have been the two jets. We couldn't prove the UFO's were the jets, but neither could we prove they weren't.

The controversial study of the UFO's' motions that Major Fournet had presented was discarded. All of the panel agreed that if there had been some permanent record of the motion of the UFO's, a photograph of a UFO's flight path or a photograph of a UFO's track on a radarscope, they could have given the study much more weight. But in every one of the ten or twenty reports that were offered as proof that the UFO's were intelligently controlled, the motions were only those that the observer had seen. And the human eye and mind are not accurate recorders. How many different stories do you get when a group of people watch two cars collide at an intersection?

Each of the fifty of our best sightings that we gave the scientists to study had some kind of a loophole. In many cases the loopholes were extremely small, but scientific evaluation has no room for even the smallest of loopholes and we had asked for a scientific evaluation.

When they had finished commenting on the reports, the scientists pointed out the seriousness of the decision they had been asked to make. They said that they had tried hard to be objective and not to be picayunish, but actually all we had was circumstantial evidence. Good circumstantial evidence, to be sure, but we had nothing concrete, no hardware, no photos showing any detail of a UFO, no measured speeds, altitudes, or sizes—nothing in the way of good, hard, cold, scientific facts. To stake the future course of millions of lives on a decision based upon circumstantial evidence would be one of the gravest mistakes in the history of the world.

In their conclusions they touched upon the possibility that the UFO's might be some type of new or yet undiscovered natural phenomenon. They explained that they hadn't given this too much credence; however, if the UFO's were a new natural phenomenon, the reports of their general appearance should follow a definite pattern— the UFO reports didn't.

This ended the section of the panel's report that covered their conclusions. The next section was entitled, "Recommendations." I fully expected that they would recommend that we as least reduce the activities of Project Blue Book if not cancel it entirely. I didn't like this one bit because I was firmly convinced that we didn't have the final answer. We needed more and better proof before a final yes or no could be given.

The panel didn't recommend that the activities of Blue Book be cut back, and they didn't recommend that it be dropped. They recommended that it be expanded. Too many of the reports had been made by credible observers, the report said, people who should know what they're looking at—people who think things out carefully. Data that was out of the circumstantial-evidence class was badly needed. And the panel must have been at least partially convinced that an expanded effort would prove something interesting because the expansion they recommended would require a considerable sum of money. The investigative force of Project Blue Book should be quadrupled in size, they wrote, and it should be staffed by specially trained experts in the fields of electronics, meteorology, photography, physics, and other fields of science pertinent to UFO investigations. Every effort should be made to set up instruments in locations where UFO sightings are frequent, so that data could be measured and recorded during a sighting. In other locations around the country military and civilian scientists should be alerted and instructed to use every piece of available equipment that could be used to track UFO's.

And lastly, they said that the American public should be told every detail of every phase of the UFO investigation—the details of the sightings, the official conclusions, and why the conclusions were made. This would serve a double purpose; it would dispel any of the mystery that security breeds and it would keep the Air Force on the ball—sloppy investigations and analyses would never occur.

When the panel's conclusions were made known in the government, they met with mixed reactions. Some people were satisfied, but others weren't. Even the opinions of a group of the country's top scientists couldn't overcome the controversy that had dogged the UFO for five years. Some of those who didn't like the decision had sat in on the UFO's trial as spectators and they felt that the "jury" was definitely prejudiced— afraid to stick their necks out. They could see no reason to continue to assume that the UFO's weren't interplanetary vehicles.

Chapter Seventeen - What Are UFO's?

While the scientists were in Washington, D.C., pondering over the UFO, the UFO's weren't just sitting idly by waiting to find out what they were— they were out doing a little "lobbying" for the cause— keeping the interest stirred up.

And they were doing a good job, too.

It was just a few minutes before midnight on January 28, 1953, when a message flashed into Wright-Patterson for Project Blue Book. It was sent "Operational Immediate," so it had priority handling; I was reading it by 12:30A.M. A pilot had chased a UFO.

The report didn't have many details but it did sound good. It gave the pilot's name and said that he could be reached at Moody AFB. I put in a long-distance call, found the pilot, and flipped on my recorder so that I could get his story word for word.

He told me that he had been flying an F-86 on a "round-robin" navigation flight from Moody AFB to Lawson AFB to Robins AFB, then back to Moody— all in Georgia. At exactly nine thirty-five he was at 6,000 feet, heading toward Lawson AFB on the first leg of his flight. He remembered that he had just looked down and had seen the lights of Albany, Georgia; then he'd looked up again and seen this bright white light at "ten o'clock high." It was an unusually bright light, and he said that he thought this was why it was so noticeable among the stars. He flew on for a few minutes watching it as he passed over Albany. He decided that it must be an extremely bright star or another airplane—except it just didn't look right. It had too much of a definitely circular shape.

It was a nice night to fly and he had to get in so much time anyway, so he thought he'd try to get a little closer to it. If it was an airplane, chances were he could close in and if it was a star, he should be able to climb up to 30,000 feet and the light shouldn't change its relative position. He checked his oxygen supply, increased the r.p.m. of the engine, and started to climb. In three or four minutes it was obvious that he was getting above the light, and he watched it; it had moved in relation to the stars. It must be an airplane then, he'd decided—an airplane so far away that he couldn't see its red and green wing tip lights.

Since he'd gone this far, he decided that he'd get closer and make sure it *was* an airplane; so he dropped the nose of the F-86 and started down. As the needle on the machmeter nudged the red line, he saw that he was getting closer because the light was getting bigger, but still he couldn't see any lights other than the one big white one. Then it wasn't white any longer; it was changing color. In about a two-second cycle it changed from white to red, then back to white again. It went through this cycle two or three times, and then before he could realize what was going on, he told me, the light changed in shape to a perfect triangle. Then it split into two triangles, one above the other. By this time he had leveled off and wasn't closing in any more. In a flash the whole thing was gone. He used the old standard description for a disappearing UFO: "It was just like someone turning off a light—it's there, then it's gone."

I asked him what he thought he'd seen. He'd thought about flying saucers, he said, but he "just couldn't swallow those stories." He thought he had a case of vertigo and the more he thought about it, the surer he was that this was the answer. He'd felt pretty foolish, he told me, and he was glad that he was alone.

Up ahead he saw the sprawling lights of Fort Benning and Lawson AFB, his turning point on the flight, and he'd started to turn but then he'd checked his fuel. The climb had used up quite a bit, so he changed his mind about going to Robins AFB and started straight back to Moody.

He called in to the ground station to change his flight plan, but before he could say anything the ground radio operator asked him if he'd seen a mysterious light.

Well—he'd seen a light.

Then the ground operator proceeded to tell him that the UFO chase had been watched on radar. First the radar had the UFO target on the scope, and it was a UFO because it was traveling much too slowly to be an airplane. Then the radar operators saw the F-86 approach, climb, and make a shallow dive toward the UFO. At first the F-86 had closed in on the UFO, but then the UFO had speeded up just enough to maintain a comfortable lead. This went on for two or three minutes; then it had moved off the scope at a terrific speed. The radar site had tried to call him, the ground station told the F-86 pilot, but they couldn't raise him so the message had to be relayed through the tower.

Rack up two more points for the UFO—another unknown and another confirmed believer.

Two or three weeks after the meeting of the panel of scientists in Washington I received word that Project Blue Book would follow the recommendations that the panel had made. I was to start implementing the plan right away. Our proposal for setting up instruments had gone to the Pentagon weeks before, so that was already taken care of. We needed more people, so I drew up a new organizational cable that called for more investigators and analysts and sent it through to ATIC's personnel section.

About this time in the history of the UFO the first of a series of snags came up. The scientists had strongly recommended that we hold nothing back— give the public everything. Accordingly, when the press got wind of the Tremonton Movie, which up until this time had been a closely guarded secret, I agreed to release it for the newsmen to see. I wrote a press release which was O.K.'d by General Garland, then the chief of ATIC, and sent it to the Pentagon. It told what the panel had said about the movies, "until proved otherwise there is no reason why the UFO's couldn't have been sea gulls." Then the release went on to say that we weren't sure exactly what the UFO's were, the sea gull theory was only an opinion. When the Pentagon got the draft of the release they screamed, "No!" No movie for the press and no press release. The sea gull theory was too weak, and we had a new publicity policy as of now—don't say anything.

This policy, incidentally, is still in effect. The January 7, 1955, issue of the *Air Force Information Services Letter* said, in essence, people in the Air Force are talking too much about UFO's— shut up. The old theory that if you ignore them they'll go away is again being followed.

Inside of a month the UFO project took a few more hard jolts. In December of 1952 I'd asked for a transfer. I'd agreed to stay on as chief of Blue Book until the end of February so that a replacement could be obtained and be broken in. But no replacement showed up. And none showed up when Lieutenant Rothstien's tour of active duty ended, when Lieutenant Andy Flues transferred to the Alaskan Air Command, or when others left. When I left the UFO project for a two-month tour of temporary duty in Denver, Lieutenant Bob Olsson took over as chief. His staff consisted of Airman First Class Max Futch. Both men were old veterans of the UFO campaign of '52, but two people can do only so much.

When I came back to ATIC in July 1953 and took over another job, Lieutenant Olsson was just getting out of the Air Force and Max Futch was now it. He said that he felt like the President of Antarctica on a non-expedition year. In a few days I again had Project Blue Book, as an additional duty this time, and I had orders to "build it up."

While I had been gone, our instrumentation plan had been rejected. Higher headquarters had decided against establishing a net of manned tracking stations, astronomical cameras tied in with radars, and our other proposed instrumentation. General Garland had argued long and hard for the plan, but he'd lost. It was decided that the cameras with diffraction gratings over the lenses, the cameras that had been under development for a year, would suffice.

The camera program had started out as a top-priority project, but it had lost momentum fast when we'd tested these widely publicized instruments and found that they wouldn't satisfactorily photograph a million-candle power flare at 450 yards. The cameras themselves were all right, but in combination with the gratings, they were no good. However, Lieutenant Olsson had been told to send them out, so he sent them out.

The first thing that I did when I returned to Project Blue Book was to go over the reports that had come in while I was away. There were several good reports but only one that was exceptional. It had taken place at Luke AFB, Arizona, the Air Force's advanced fighter-bomber school that is named after the famous "balloon buster" of World War I, Lieutenant Frank Luke, Jr. It was a sighting that produced some very interesting photographs.

There were only a few high cirrus clouds in the sky late on the morning of March 3 when a pilot took off from Luke in an F-84 jet to log some time. He had been flying F-51's in Korea and had recently started to check out in the jets. He took off, cleared the traffic pattern, and started climbing toward Blythe Radio, about 130 miles west of Luke. He'd climbed for several minutes and had just picked up the coded letters BLH that identified Blythe Radio when he looked up through the corner glass in the front part of his canopy— high at about two o'clock he saw what he thought was an airplane angling across his course from left to right leaving a long, thin vapor trail. He glanced down at his altimeter and saw that he was at 23,000 feet. The object that was leaving the vapor trail must really be high, he remembered thinking, because he couldn't see any airplane at the head of it. He altered his course a few degrees to the right so that he could follow the trail and increased his rate of climb. Before long he could tell that he was gaining on the object, or whatever was leaving the vapor trail, because he was under the central part of it. But he still couldn't see any object. This was odd, he thought, because vapor trails don't just happen; something has to leave them. His altimeter had ticked off another 12,000 feet and he was now at 35,000. He kept on climbing, but soon the '84 began to mush; it was as high as it would go. The pilot dropped down 1,000 feet and continued on—now he was below the front of the trail, but still no airplane. This bothered him too. Nothing that we have flies over 55,000 feet except a few experimental airplanes like the D-558 or those of the "X" series, and they don't stray far from Edwards AFB in California. He couldn't be more than 15,000 feet from the front of the trail, and you can recognize any kind of an airplane 15,000 feet away in the clear air of the substratosphere. He looked and he looked and he looked. He rocked the F-84 back and forth thinking maybe he had a flaw in the plexiglass of the canopy that was blinking out the airplane, but still no airplane. Whatever it was, it was darn high or darn small. It was moving about 300 miles an hour because he had to pull off power and "S" to stay under it.

He was beginning to get low on fuel about this time so he hauled up the nose of the jet, took about 30 feet of gun camera film, and started down. When he landed and told his story, the film was quickly processed and rushed to the projection room. It showed a weird, thin, forked vapor trail— but no airplane.

Lieutenant Olsson and Airman Futch had worked this one over thoroughly. The photo lab confirmed that the trail was definitely a vapor trail, not a freak cloud formation. But Air Force Flight Service said, "No other airplanes in the area," and so did Air Defense Command, because minutes after the F-84 pilot broke off contact, the "object" had passed into an ADIZ—Air Defense Identification Zone—and radar had shown nothing.

There was one last possibility: Blue Book's astronomer said that the photos looked exactly like a meteor's smoke trail. But there was one hitch: the pilot was positive that the head of the vapor trail was moving at about 300 miles an hour. He didn't know exactly how much ground he'd covered, but when he first picked up Blythe Radio he was on Green 5 airway, about 30 miles west of his base, and when he'd given up the chase he'd gotten another radio bearing, and he was now almost up to Needles Radio, 70 miles north of Blythe. He could see a lake, Lake Mojave, in the distance.

Could a high-altitude jet-stream wind have been blowing the smoke cloud? Futch had checked this—no. The winds above 20,000 feet were the usual westerlies and the jet stream was far to the north.

Several months later I talked to a captain who had been at Luke when this sighting occurred. He knew the F-84 pilot and he'd heard him tell his story in great detail. I won't say that he was a confirmed believer, but he was interested. "I never thought much about these reports before," he said, "but I know this guy well. He's not nuts. What do you think he saw?"

I don't know what he saw. Maybe he didn't travel as far as he thought he did. If he didn't, then I'd guess that he saw a meteor's smoke trail. But if he did know that he'd covered some 80 miles during the chase, I'd say that he saw a UFO—a real one. And I find it hard to believe that pilots don't know what they're doing.

During the summer of 1953, UFO reports dropped off considerably. During May, June, and July of 1952 we'd received 637 good reports. During the same months in 1953 we received only seventy-six. We had been waiting for the magic month of July to roll around again because every July there had been the sudden and unexplained peak in reporting; we wanted to know if it would happen again. It didn't— only twenty-one reports came in, to make July the lowest month of the year. But July did bring new developments.

Project Blue Book got a badly needed shot in the arm when an unpublicized but highly important change took place: another intelligence agency began to take over all field investigations.

Ever since I'd returned to the project, the orders had been to build it up— get more people—do what the panel recommended. But when I'd asked for more people, all I got was a polite "So sorry." So, I did the next best thing and tried to find some organization already in being which could and would help us. I happened to be expounding my troubles one day at Air Defense Command Headquarters while I was briefing General Burgess, ADC's Director of Intelligence, and he told me about his 4602nd Air Intelligence Squadron, a specialized intelligence unit that had recently become operational. Maybe it could help—he'd see what he could work out, he told me.

Now in the military all commitments to do something carry an almost standard time factor. "I'll expedite it," means nothing will happen for at least two weeks. "I'll do it right away," means from a month to six weeks. An answer like, "I'll see what I can work out," requires writing a memo that explains what the person was going to see if he could work out and sealing it in a time capsule for preservation so that when the answer finally does come through the future generation that receives it will know how it all started. But I underestimated the efficiency of the Air Defense Command. Inside of two weeks General Burgess had called General Garland, they'd discussed the problem, and I was back in Colorado Springs setting up a program with Colonel White's 4602nd.

The 4602nd's primary function is to interrogate captured enemy airmen during wartime; in peacetime all that they can do is participate in simulated problems. Investigating UFO reports would supplement these problems and add a factor of realism that would be invaluable in their training. The 4602nd had field teams spread out all over the United States, and these teams could travel anywhere by airplane, helicopter, canoe, jeep, or skis on a minute's notice. The field teams had already established a working contact with the highway patrols, sheriffs' offices, police, and the other military in their respective areas, so they were in an excellent position to collect facts about a UFO report. Each member of the field teams had been especially chosen and trained in the art of interrogation, and each team had a technical specialist. We couldn't have asked for a better ally.

Project Blue Book was once more back in business. Until the formal paper work went through, our plan was that whenever a UFO report worth investigating came in we would call the 4602nd and they would get a team out right away. The team would make a thorough investigation and wire us their report. If the answer came back "Unknown," we would study the details of the sighting and, with the help of Project Bear, try to find the answer.

A few weeks after the final plans had been made with the 4602nd, I again bade farewell to Project Blue Book. In a simple ceremony on the poop deck of one of the flying saucers that I frequently have been accused of capturing, before a formation of the three-foot-tall green men that I have equally as frequently been accused of keeping prisoner, I turned my command over to Al/c Max Futch and walked out the door into civilian life with separation orders in hand.

The UFO's must have known that I was leaving because the day I found out that officers with my specialty, technical intelligence, were no longer on the critical list and that I could soon get out of the service, they really put on a show. The show they put on is still the best UFO report in the Air Force files.

I first heard about the sighting about two o'clock on the morning of August 13, 1953, when Max Futch called me from ATIC. A few minutes before a wire had come in carrying a priority just under that reserved for flashing the word the U.S. has been attacked. Max had been called over to ATIC by the OD to see the report, and he thought that I should see it. I was a little hesitant to get dressed and go out to the base, so I asked Max what he thought about the report. His classic answer will go down in UFO history, "Captain," Max said in his slow, pure Louisiana drawl, "you know that for a year I've read every flying saucer report that's come in and that I never really believed in the things." Then he hesitated and added, so fast that I could hardly understand him, "But you should read *this* wire." The speed with which he uttered this last statement was in itself enough to convince me. When Max talked fast, something was important.

A half hour later I was at ATIC—just in time to get a call from the Pentagon. Someone else had gotten out of bed to read his copy of the wire. I used the emergency orders that I always kept in my desk and caught the first airliner out of Dayton to Rapid City, South Dakota. I didn't call the 4602nd because I wanted to investigate this one personally. I talked to everyone involved in the incident and pieced together an amazing story.

Shortly after dark on the night of the twelfth, the Air Defense Command radar station at Ellsworth AFB, just east of Rapid City, had received a call from the local Ground Observer Corps filter center. A lady spotter at Black Hawk, about 10 miles west of Ellsworth, had reported an extremely bright light low on the horizon, off to the northeast. The radar had been scanning an area to the west, working a jet fighter in some practice patrols, but when they got the report they moved the sector scan to the northeast quadrant. There was a target exactly where the lady reported the light to be. The warrant officer, who was the duty controller for the night, told me that he'd studied the target for several minutes. He knew how weather could affect radar but this target was "well defined, solid, and bright." It seemed to be moving, but very slowly. He called for an altitude reading, and the man on the height-finding radar checked his scope. He also had the target—it was at 16,000 feet.

The warrant officer picked up the phone and asked the filter center to connect him with the spotter. They did, and the two people compared notes on the UFO's position for several minutes. But right in the middle of a sentence the lady suddenly stopped and excitedly said, "It's starting to move—it's moving southwest toward Rapid."

The controller looked down at his scope and the target was beginning to pick up speed and move southwest. He yelled at two of his men to run outside and take a look. In a second or two one of them shouted back that they could both see a large bluish-white light moving toward Rapid City. The controller looked down at his scope—the target was moving toward Rapid City. As all three parties watched the light and kept up a steady cross conversation of the description, the UFO swiftly made a wide sweep around Rapid City and returned to its original position in the sky.

A master sergeant who had seen and heard the happenings told me that in all his years of duty—combat radar operations in both Europe and Korea—he'd never been so completely awed by anything. When the warrant officer had yelled down at him and asked him what he thought they should do, he'd just stood there. "After all," he told me, "what in hell could we do—they're bigger than all of us."

But the warrant officer did do something. He called to the F-84 pilot he had on combat air patrol west of the base and told him to get ready for an intercept. He brought the pilot around south of the base and gave him a course correction that would take him right into the light, which was still at 16,000 feet. By this time the pilot had it spotted. He made the turn, and when he closed to within about 3 miles of the target, it began to move. The controller saw it begin to move, the spotter saw it begin to move and the pilot saw it begin to move—all at the same time. There was now no doubt that all of them were watching the same object.

Once it began to move, the UFO picked up speed fast and started to climb, heading north, but the F-84 was right on its tail. The pilot would notice that the light was getting brighter, and he'd call the controller to tell him about it. But the controller's answer would always be the same, "Roger, we can see it on the scope."

There was always a limit as to how near the jet could get, however. The controller told me that it was just as if the UFO had some kind of an automatic warning radar linked to its power supply. When something got too close to it, it would automatically pick up speed and pull away. The separation distance always remained about 3 miles.

The chase continued on north—out of sight of the lights of Rapid City and the base—into some very black night.

When the UFO and the F-84 got about 120 miles to the north, the pilot checked his fuel; he had to come back. And when I talked to him, he said he was damn glad that he was running out of fuel because being out over some mighty desolate country alone with a UFO can cause some worry.

Both the UFO and the F-84 had gone off the scope, but in a few minutes the jet was back on, heading for home. Then 10 or 15 miles behind it was the UFO target also coming back.

While the UFO and the F-84 were returning to the base—the F-84 was planning to land—the controller received a call from the jet interceptor squadron on the base. The alert pilots at the squadron had heard the conversations on their radio and didn't believe it. "Who's nuts up there?" was the comment that passed over the wire from the pilots to the radar people. There was an F-84 on the line ready to scramble, the man on the phone said, and one of the pilots, a World War II and Korean veteran, wanted to go up and see a flying saucer. The controller said, "O.K., go."

In a minute or two the F-84 was airborne and the controller was working him toward the light. The pilot saw it right away and closed in. Again the light began to climb out, this time more toward the northeast. The pilot also began to climb, and before long the light, which at first had been about 30 degrees above his horizontal line of sight, was now below him. He nosed the '84 down to pick up speed, but it was the same old story—as soon as he'd get within 3 miles of the UFO, it would put on a burst of speed and stay out ahead.

Even though the pilot could see the light and hear the ground controller telling him that he was above it, and alternately gaining on it or dropping back, he still couldn't believe it—there must be a simple explanation. He turned off all of his lights—it wasn't a reflection from any of the airplane's lights because there it was. A reflection from a ground light, maybe. He rolled the airplane—the position of the light didn't change. A star—he picked out three bright stars near the light and watched carefully. The UFO moved in relation to the three stars. Well, he thought to himself, if it's a real object out there, my radar should pick it up too; so he flipped on his radar-ranging gunsight. In a few seconds the red light on his sight blinked on—something real and solid was in front of him. Then he was scared. When I talked to him, he readily admitted that he'd been scared. He'd met MD 109's, FW 190's and ME 262's over Germany and he'd met MIG-15's over Korea but the large, bright, bluish-white light had scared him—he asked the controller if he could break off the intercept.

This time the light didn't come back.

When the UFO went off the scope it was headed toward Fargo, North Dakota, so the controller called the Fargo filter center. "Had they had any reports of unidentified lights?" he asked. They hadn't.

But in a few minutes a call came back. Spotter posts on a southwest-northeast line a few miles west of Fargo had reported a fast-moving, bright bluish-white light.

This was an unknown—the best.

The sighting was thoroughly investigated, and I could devote pages of detail on how we looked into every facet of the incident; but it will suffice to say that in every facet we looked into we saw nothing. Nothing but a big question mark asking what was it.

When I left Project Blue Book and the Air Force I severed all official associations with the UFO. But the UFO is like hard drink; you always seem to drift back to it. People I've met, people at work, and friends of friends are continually asking about the subject. In the past few months the circulation manager of a large Los Angeles newspaper, one of Douglas Aircraft Company's top scientists, a man who is guiding the future development of the supersecret Atlas intercontinental guided missile, a movie star, and a German rocket expert have called me and wanted to get together to talk about UFO's. Some of them had seen one.

I have kept up with the activity of the UFO and Project Blue Book over the past two years through friends who are still in intelligence. Before Max Futch got out of the Air Force and went back to law school he wrote to me quite often and a part of his letters were always devoted to the latest about the UFO's.

Then I make frequent business trips to ATIC, and I always stop in to see Captain Charles Hardin, who is now in charge of Blue Book, for a "What's new?" I always go to ATIC with the proper security clearances so I'm sure I get a straight answer to my question.

Since I left ATIC, the UFO's haven't gone away and neither has the interest. There hasn't been too much about them in the newspapers because of the present Air Force policy of silence, but they're with us. That the interest is still with us is attested to by the fact that in late 1953 Donald Keyhoe's book about UFO's, *Flying Saucers from Outer Space*, immediately appeared on best seller lists. The book was based on a few of our good UFO reports that were released to the press. To say that the book is factual depends entirely upon how one uses the word. The details of the specific UFO sightings that he credits to the Air Force are factual, but in his interpretations of the incidents he blasts way out into the wild blue yonder.

During the past two years the bulk of the UFO activity has taken place in Europe. I might add here that I have never seen any recent official UFO reports or studies from other countries; all of my information about the European Flap came from friends. But when these friends are in the intelligence branches of the U.S. Air Force, the RAF, and the Royal Netherlands Air Force, the data can be considered at least good.

The European Flap started in the summer of 1953, when reports began to pop up in England and France. Quality-wise these first reports weren't too good, however. But then, like a few reports that occurred early in the stateside Big Flap of 1952, sightings began to drift in that packed a bit of a jolt. Reports came in that had been made by personal friends of the brass in the British and French Air Forces. Then some of the brass saw them. Corners of mouths started down.

In September several radar sites in the London area picked up unidentified targets streaking across the city at altitudes of from 44,000 to 68,000 feet. The crews who saw the targets said, "Not weather," and some of these crews had been through the bloody Battle of Britain. They knew their radar.

In October the crew of a British European Airways airliner reported that a "strange aerial object" had paced their twin-engined Elizabethan for thirty minutes. Then on November 3, about two-thirty in the afternoon, radar in the London area again picked up targets. This time two Vampire jets were scrambled and the pilots saw a "strange aerial object." The men at the radar site saw it too; through their telescope it looked like a "flat, white-coloured tennis ball."

The flap continued into 1954. In January those people who officially keep track of the UFO's pricked up their ears when the report of two Swedish airline pilots came in. The pilots had gotten a good look before the UFO had streaked into a cloud bank. It looked like a discus with a hump in the middle.

On through the spring reports poured out of every country in Europe. Some were bad, some were good.

On July 3, 1954, at eight-fifteen in the morning, the captain, the officers and 463 passengers on a Dutch ocean liner watched a "greenish-colored, saucer-shaped object about half the size of a full moon" as it sped across the sky and disappeared into a patch of high clouds.

There was one fully documented and substantiated case of a "landing" during the flap. On August 25 two young ladies in Mosjoen, Norway, made every major newspaper in the world when they encountered a "saucer-man." They said that they were picking berries when suddenly a dark man, with long shaggy hair, stepped out from behind some bushes. He was friendly; he stepped right up to them and started to talk rapidly. The two young ladies could understand English but they couldn't understand him. At first they were frightened, but his smile soon "disarmed" them. He drew a few pictures of flying saucers and pointed up in the sky. "He was obviously trying to make a point," one of the young ladies said.

A few days later it was discovered that the man from "outer space" was a lost USAF helicopter pilot who was flying with NATO forces in Norway.

As I've always said, "Ya gotta watch those Air Force pilots— especially those shaggy-haired ones from Brooklyn."

The reporting spread to Italy, where thousands of people in Rome saw a strange cigar-shaped object hang over the city for forty minutes. Newspapers claimed that Italian Air Force radar had the UFO on their scopes, but as far as I could determine, this was never officially acknowledged.

In December a photograph of two UFO's over Taormina, Sicily, appeared in many newspapers. The picture showed three men standing on a bridge, with a fourth running up with a camera. All were intently watching two disk-shaped objects. The photo looked good, but there was one flaw, the men weren't looking at the UFO's; they were looking off to the right of them. I'm inclined to agree with Captain Hardin of Blue Book—the photographer just fouled up on his double exposure.

Sightings spread across southern Europe, and at the end of October, the Yugoslav Government expressed official interest. Belgrade newspapers said that a "thoughtful inquiry" would be set up, since reports had come from "control tower operators, weather stations and hundreds of farmers." But the part of the statement that swung the most weight was, "Scientists in astronomical observatories have seen these strange objects with their own eyes."

During 1954 and the early part of 1955 my friends in Europe tried to keep me up-to-date on all of the better reports, but this soon approached a full-time job. Airline pilots saw them, radar picked them up, and military pilots chased them. The press took sides, and the controversy that had plagued the U.S. since 1947 bloomed forth in all its confusion.

An ex-Air Chief Marshal in the RAF, Lord Dowding, went to bat for the UFO's. The Netherlands Air Chief of Staff said they can't be. Herman Oberth, the father of the German rocket development, said that the UFO's were definitely interplanetary vehicles.

In Belgium a senator put the screws on the Secretary of Defense—he wanted an answer. The Secretary of Defense questioned the idea that the saucers were "real" and said that the military wasn't officially interested. In France a member of parliament received a different answer—the French military was interested. The French General Staff had set up a committee to study UFO reports.

In Italy, Clare Boothe Luce, American Ambassador to Italy, said that she had seen a UFO and had no idea what it could be.

Halfway around the world, in Australia, the UFO's were busy too. At Canberra Airport the pilot of an RAAF Hawker Sea Fury and a ground radar station teamed up to get enough data to make an excellent radar- visual report.

In early 1955 the flap began to die down about as rapidly as it had flared up, but it had left its mark—many more believers. Even the highly respected British aviation magazine, *Aeroplane*, had something to say. One of the editors took a long, hard look at the over-all UFO picture and concluded, "Really, old chaps—I don't know."

Probably the most unique part of the whole European Flap was the fact that the Iron Curtain countries were having their own private flap. The first indications came in October 1954, when Rumanian newspapers blamed the United States for launching a drive to induce a "flying saucer psychosis" in their country. The next month the Hungarian Government hauled an "expert" up in front of the microphone so that he could explain to the populace that UFO's don't really exist because, "all 'flying saucer' reports originate in the bourgeois countries, where they are invented by the capitalist warmongers with a view to drawing the people's attention away from their economic difficulties."

Next the U.S.S.R. itself took up the cry along the same lines when the voice of the Soviet Army, the newspaper *Red Star*, denounced the UFO's as, you guessed it, capitalist propaganda.

In 1955 the UFO's were still there because the day before the all- important May Day celebration, a day when the Soviet radio and TV are normally crammed with programs plugging the glory of Mother Russia to get the peasants in the mood for the next day, a member of the Soviet Academy of Sciences had to get on the air to calm the people's fears. He left out Wall Street and Dulles this time—UFO's just don't exist.

It was interesting to note that during the whole Iron Curtain Flap, not one sighting or complimentary comment about the UFO's was made over the radio or in the newspapers; yet the flap continued. The reports were obviously being passed on by word of mouth. This fact seems to negate the theory that if the newspaper reporters and newscasters would give up the UFO's would go away. The people in Russia were obviously seeing something.

While the European Flap was in progress, the UFO's weren't entirely neglecting the United States. The number of reports that were coming into Project Blue Book were below average, but there were reports. Many of them would definitely be classed as good, but the best was a report from a photo reconnaissance B-29 crew that encountered a UFO almost over Dayton.

About 11:00A.M. on May 24, 1954, an RB-29 equipped with some new aerial cameras took off from Wright Field, one of the two airfields that make up Wright-Patterson AFB, and headed toward the Air Force's photographic test range in Indiana. At exactly twelve noon they were at 16,000 feet, flying west, about 15 miles northwest of Dayton. A major, a photo officer, was in the nose seat of the '29. All of the gun sights and the bombsight in the nose had been taken out, so it was like sitting in a large picture window—except you just can't get this kind of a view anyplace else. The major was enjoying it. He was leaning forward, looking down, when he saw an extremely bright circular-shaped object under and a little behind the airplane. It was so bright that it seemed to have a mirror finish. He couldn't tell how far below him it was but he was sure that it wasn't any higher than 6,000 feet above the ground, and it was traveling fast, faster than the B-29. It took only about six seconds to cross a section of land, which meant that it was going about 600 miles an hour.

The major called the crew and told them about the UFO, but neither the pilot nor the copilot could see it because it was now directly under the B-29. The pilot was just in the process of telling him that he was crazy when one of the scanners in an aft blister called in; he and the other scanner could also see the UFO.

Being a photo ship, the RB-29 had cameras—loaded cameras—so the logical thing to do would be to take a picture, but during a UFO sighting logic sometimes gets shoved into the background. In this case, however, it didn't, and the major reached down, punched the button on the intervalometer, and the big vertical camera in the aft section of the airplane clicked off a photo before the UFO sped away.

The photo showed a circular-shaped blob of light exactly as the major had described it to the RB-29 crew. It didn't show any details of the UFO because the UFO was too bright; it was completely overexposed on the negative. The circular shape wasn't sharp either; it had fuzzy edges, but this could have been due to two things: its extreme brightness, or the fact that it was high, close to the RB-29, and out of focus. There was no way of telling exactly how high it was but if it were at 6,000 feet, as the major estimated, it would have been about 125 feet in diameter.

Working with people from the photo lab at Wright-Patterson, Captain Hardin from Project Blue Book carried out one of the most complete investigations in UFO history. They checked aircraft flights, rephotographed the area from high and low altitude to see if they could pick up something on the ground that could have been reflecting light, and made a minute ground search of the area. They found absolutely nothing that could explain the round blob of light, and the incident went down as an unknown.

Like all good "Unknown" UFO reports, there are as many opinions as to what the bright blob of light could have been as there are people who've seen the photo. "Some kind of light phenomenon" is the frequent opinion of those who don't believe. They point out that there is no shadow of any kind of a circular object showing on the ground—no shadow, nothing "solid." But if you care to take the time you can show that if the object, assuming that this is what it was, was above 4,000 feet the shadow would fall out of the picture.

Then all you get is a blank look from the light phenomenon theorists.

With the sighting from the RB-29 and the photograph, all of the other UFO reports that Blue Book has collected and all of those that came out of the European Flap, the big question—the key question— is: What have the last two years of UFO activity brought out? Have there been any important developments?

Some good reports have come in and the Air Force is sitting on them. During 1954 they received some 450 reports, and once again July was the peak month. In the first half of 1955 they had 189. But I can assure you that these reports add nothing more as far as proof is concerned. The quality of the reports has improved, but they still offer nothing more than the same circumstantial evidence that we presented to the panel of scientists in early 1953. There have been no reports in which the speed or altitude of a UFO has been measured, there have been no reliable photographs that show any details of a UFO, and there is no hardware. There is still no real proof.

So a public statement that was made in 1952 still holds true: "The *possibility* of the existence of interplanetary craft has never been denied by the Air Force, *but* UFO reports offer absolutely no authentic evidence that such interplanetary spacecraft do exist."

But with the UFO, what is lacking in proof is always made up for in opinions. To get a qualified opinion, I wrote to a friend, Frederick C. Durant. Mr. Durant, who is presently the director of a large Army Ordnance test station, is also a past president of the American Rocket Society and president of the International Astronautical Federation. For those who are not familiar with these organizations, the American Rocket Society is an organization established to promote interest and research in space flight and lists as its members practically every prominent scientist and engineer in the professional fields allied to aeronautics. The International Astronautical Federation is a world-wide federation of such societies.

Mr. Durant has spent many hours studying UFO reports in the Project Blue Book files and many more hours discussing them with scientists the world over—scientists who are doing research and formulating the plans for space flight. I asked him what he'd heard about the UFO's during the past several years and what he thought about them. This was his reply:

This past summer at the Annual Congress of the IAF at Innsbruck, as well as previous Congresses (Zurich, 1953, Stuttgart, 1952, and London, 1951), none of the delegates representing the rocket and space flight societies of all the countries involved had strong feelings on the subject of saucers. Their attitude was essentially the same as professional members of the American Rocket Society in this country. In other words, there appear to be no confirmed saucer fans in the hierarchy of the professional societies.

I continue to follow the subject of UFO's primarily because of my being requested for comment on the interplanetary flight aspects. My personal feelings have not changed in the past four years, although I continue to keep an objective outlook.

There are many other prominent scientists in the world whom I met while I was chief of Project Blue Book who, I'm sure, would give the same answer—they've not been able to find any proof, but they continue to keep an objective outlook. There are just enough big question marks sprinkled through the reports to keep their outlook objective.

I know that there are many other scientists in the world who, although they haven't studied the Air Force's UFO files, would limit their comment to a large laugh followed by an "It can't be." But "It can't be's" are dangerous, if for no other reason than history has proved them so.

Not more than a hundred years ago two members of the French Academy of Sciences were unseated because they supported the idea that "stones had fallen from the sky." Other distinguished members of the French Academy examined the stones, "It can't be—stones don't fall from the sky," or words to that effect. "These are common rocks that have been struck by lightning."

Today we know that the "stones from the sky" were meteorites.

Not more than fifty years ago Dr. Simon Newcomb, a world-famous astronomer and the first American since Benjamin Franklin to be made an associate of the Institute of France, the hierarchy of the world science, said, "It can't be." Then he went on to explain that flight without gas bags would require the discovery of some new material or a new force in nature.

And at the same time Rear Admiral George W. Melville, then Chief Engineer for the U.S. Navy, said that attempts to fly heavier-than- air vehicles was absurd.

Just a little over ten years ago there was another "it can't be." Ex- President Harry S. Truman recalls in the first volume of the Truman *Memoirs* what Admiral William D. Leahy, then Chief of Staff to the President, had to say about the atomic bomb. "That is the biggest fool thing we have ever done," he is quoted as saying. "The bomb will never go off, and I speak as an expert in explosives."

Personally, I don't believe that "it can't be." I wouldn't class myself as a "believer," exactly, because I've seen too many UFO reports that first appeared to be unexplainable fall to pieces when they were thoroughly investigated. But every time I begin to get skeptical I think of the other reports, the many reports made by experienced pilots and radar operators, scientists, and other people who know what they're looking at. These reports were thoroughly investigated and they are still unknowns. Of these reports, the radar- visual sightings are the most convincing. When a ground radar picks up a UFO target and a ground observer sees a light where the radar target is located, then a jet interceptor is scrambled to intercept the UFO and the pilot also sees the light and gets a radar lock-on only to have the UFO almost impudently outdistance him, there is no simple answer. We have no aircraft on this earth that can at will so handily outdistance our latest jets.

The Air Force is still actively engaged in investigating UFO reports, although during the past six months there have been definite indications that there is a movement afoot to get Project Blue Book to swing back to the old Project Grudge philosophy of analyzing UFO reports—write them all off, regardless. But good UFO reports cannot be written off with such answers as fatigued pilots seeing a balloon or star; "green" radar operators with *only* fifteen years' experience watching temperature inversion caused blips on their radarscopes; or "a mild form of mass hysteria or war nerves." Using answers like these, or similar ones, to explain the UFO reports is an expedient method of getting the percentage of unknowns down to zero, but it is no more valid than turning the hands of a clock ahead to make time pass faster. Twice before the riddle of the UFO has been "solved," only to have the reports increase in both quantity and quality.

I wouldn't want to hazard a guess as to what the final outcome of the UFO investigation will be, but I am sure that within a few years there will be a proven answer. The earth satellite program, which was recently announced, research progress in the fields of electronics, nuclear physics, astronomy, and a dozen other branches of the sciences will furnish data that will be useful to the UFO investigators. Methods of investigating and analyzing UFO reports have improved a hundredfold since 1947 and they are continuing to be improved by the diligent work of Captain Charles Hardin, the present chief of Project Blue Book, his staff, and the 4602nd Air Intelligence Squadron. Slowly but surely these people are working closer to the answer—closer to the proof.

Maybe the final proven answer will be that all of the UFO's that have been reported are merely misidentified known objects. Or maybe the many pilots, radar specialists, generals, industrialists, scientists, and the man on the street who have told me, "I wouldn't have believed it either if I hadn't seen it myself," knew what they were talking about. Maybe the earth is being visited by interplanetary spaceships.

Only time will tell.

Chapter Eighteen - And They're Still Flying

Four years have passed since the first seventeen chapters of this book were written. During this period hundreds of unidentified flying objects have been seen and reported to the Air Force. Pilots, with thousands of hours of flying time are still reporting them; radar operators, experts in their field, are still tracking them; and crews on the missile test ranges are photographing them.

UFO's are not just a fad.

The Air Force's Project Blue Book is still very active. Not a week passes that one of the many teams of its nation wide investigation net is not in the field investigating a new UFO report.

To pick up the history of the UFO the best place to start is Cincinnati, Ohio, in the late summer of 1955. For some unknown reason, one of those mysterious factors of the UFO, reports from this Hamilton County city suddenly began to pick up. Mass hysteria, the old crutch, wasn't a factor because neither the press, the radio nor TV was even mentioning the words "flying saucer."

The reports weren't much in terms of quality. Some lady would see a "bobbing white light"; or a man, putting his car away, would see a "star jump." These reports, usually passed on to the Air Force through the Air Defense Command's Ground Observer Corps, merely went on the UFO plotting board as a statistic.

But before long, in a matter of a week or two, the mass of reports began to draw some official attention because the Ground Observer Corps spotters themselves began to make UFO reports. At times during the middle of August the telephone lines from the GOC observation posts in Hamilton County (greater Cincinnati) to the filter center in Columbus would be jammed. Now, even the most cynical Air Force types were be-grudgingly raising their eyebrows. These GOC observers were about as close to "experts" as you can get. Many had spent hundreds of hours scanning the skies since the GOC went into the operation in 1952 to close the gaps in our radar net. Many held awards for meritorious service. They weren't crackpots.

But still the cynics held out. This was really nothing new. The Project Blue Book files were full of similar incidents. In 1947 there had been a rash of reports from the Pacific Northwest; in 1948 there had been a similar outbreak at Edwards Air Force Base, the supersecret test center in the Mojave Desert of California; in 1949 the sightings centered in the midwest. None had panned out to be anything.

Then came the clincher.

On the night of August 23rd, shortly before midnight, reports of a UFO began to come in from the Mt. Healthy GOC observation post northwest of Cincinnati. Almost simultaneously, Air Defense Command radar picked up a target in that area. A minute or two later the Forestville and Loveland GOC posts, also in Hamilton County, made sightings. Now, three UFO's, described as brilliant white spheres, swinging in a pendulum-like motion, were on the ADC plotting boards- confirmed by radar. All pretext of ignoring the UFO's was dropped and at 11:58P.M., F-84's of the Ohio Air National Guard were scrambled. They were over Cincinnati at 12:10A.M. and made contact. Boring in at 20,000 feet, at 100% power, they closed but the UFO's left them as if they were standing still.

The battle in the Cincinnati sector was on.

Almost every night more UFO's were reported by the GOC. Attempts were made to scramble interceptors but there were no more radar contacts and a jet interceptor without ground guidance is worthless.

At the height of this activity it was decided that more information was needed by the Air Defense Command. Maybe from a mass of data something, some kind of clue, could be sifted out. The answer: establish a special UFO reporting post. The man to operate this post was tailor-made.

On September 9, Major Hugh McKenzie of the Columbus Filter Center contacted Leonard H. Stringfield in Cincinnati. Stringfield, besides being a very public minded citizen, was also known as a level-headed "saucer expert." Sooner or later, usually sooner, he heard about every UFO sighting in Hamilton County. He was given a code, "Foxtrot Kilo 3-0 Blue," which provided him with an open telephone line to the ADC Filter Center in Columbus. He was in business but he didn't have to build up a clientele—it was there.

For the next few months Stringfield did yeoman duty as Cincinnati's one-man UFO center by sifting out the wheat from the chaff and passing the wheat on to the Air Force. As he told me the other day, half his nights were spent in his backyard clad in shorts and binoculars. Fortunately his neighbors were broad-minded and the UFO's picked relatively warm nights to appear.

Most of the reports Stringfield received were duds. He lost track of the number. The green, red, blue, gold and white; discs, triangles, squares and footballs which hovered, streaked, zigzagged and jerked, turned out to be Venus, Jupiter, Arcturus and an occasional jet. A fiery orange satellite which hovered for hours turned out to be the North Star viewed through a cheap telescope, and the "whole formation of space ships" were the Pleiades.

Then it happened again.

On the evening of March 23rd Stringfield's telephone rang. It was Charles Deininger at the Mt. Healthy GOC post. They had a UFO in sight off to the east. Could Stringfield see it? He grabbed his extension phone and ran outdoors. There, off to the east, were two, large, low flying lights. One of the lights was a glowing green and the other yellow. They were moving north.

"Airplane!"

This was Stringfield's first reaction but during World War II he had made the long trek up the Pacific with the famous Fifth Air Force and he immediately realized that if it was an airplane it would have to be very close because of the large distance between the lights. And, as a clincher, no sound came through the still night.

He dialed the long distance operator and said the magic words, "This is Foxtrot Kilo Three Dash Zero Blue." Seconds later he was talking to the duty sergeant at the Columbus Filter Center. A few more seconds and the sergeant had his story.

Another jet was scrambled and this time Stringfield, via a radiotelephone hookup to the airplane, gave the pilot a vector. Stringfield heard the jet closing in but since it was a one-way circuit he couldn't hear the pilot's comments.

Once again the UFO took off.

This was a fitting climax for the Cincinnati flap. As suddenly as it began it quit and from the mass of data that was collected the Air Force got zero information.

In the mystery league the UFO's were still ahead.

Although the majority of the UFO activity during the last half of 1955 and early 1956 centered in the Cincinnati area there were other good reports.

Near Banning, California, on November 25, 1955, Gene Miller, manager of the Banning Municipal Airport and Dr. Leslie Ward, a physician, were paced by a "globe of white light which suddenly backed up in midair," while in Miller's airplane. It was the same old story: Miller was an experienced pilot, a former Air Force instructor and air freight pilot with several thousand hours flying time.

Commercial pilots came in for more than their share of the sightings in 1956.

On January 22, UFO investigators talked to the crew of a Pan American airliner. That night, at 8:30P.M., the Houston to Miami DC-7B had been "abeam" of New Orleans, out over the Gulf of Mexico. There was a partial moon shining through small wisps of high cirrus clouds but generally it was a clear night. The captain of the flight was back in the cabin chatting with the passengers; the co-pilot and engineer were alone on the flight deck. The engineer had moved up from his control panel and was sitting beside the co-pilot.

At 8:30 it was time for a radio position report and the co-pilot, Tom Tompkins, leaned down to set up a new frequency on the radio controls. Robert Mueller, the engineer, was on watch for other aircraft. It was ten, maybe twenty seconds after Tompkins leaned down that Mueller just barely perceived a pinpoint of moving light off to his right. Even before his thought processes could tell him it might be another airplane the light began to grow in size. Within a short six seconds it streaked across the nose of the airliner, coming out of the Gulf and disappearing inland over Mississippi or Alabama. Tompkins, the co-pilot, never saw it because Mueller was too astounded to even utter a sound.

But Mueller had a good look. The body of the object was shaped like a bullet and gave off a "pale, luminescent blue glow." The stubby tail, or exhaust, was marked by "spurts of yellow flame or light."

The size? Mueller, like any experienced observer, had no idea since he didn't know how far away it was. But, it was big!

One sentence, dangling at the bottom of the report was one I'd seen many, many times before: "Mr. Mueller *was* a complete skeptic regarding UFO reports."

During 1956 there was a rumor—I heard it many times—that the Air Force had entered into a grand conspiracy with the U.S. news media to "stamp out the UFO." The common people of the world, the rumor had it, were not yet psychologically conditioned to learn that we had been visited by superior beings. By not ever mentioning the words "unidentified flying object" the public would forget and go on their merry, stupid way. I heard this rumor so often, in fact, that I began to wonder myself. But a few dollars invested in Martinis for old buddies in the Kittyhawk Room of the Biltmore Hotel in Dayton, or the Men's bar in the Statler Hotel in Washington, produces a lot of straight and reliable information—much better than you get through official channels. There was no "silence" order I learned, only the same old routine. If the files at ATIC were opened to the public it would take a staff of a dozen people to handle all the inquiries.

Secondly, many of the inquiries come from saucer screwballs and these people are like a hypochondriac at the doctor's; nothing will make them believe the diagnosis unless it is what they came in to hear. And there are plenty of saucer screwballs.

One officer summed it up neatly when he told me, "It isn't the UFO's that give us the trouble, it's the people."

As a double check I called several newspaper editors the other day and asked, "Why don't you print more UFO stories?" The answers were simple, it's the old "dog bites man" bit—ninety-nine per cent have no news value any more.

On May 10, 1956, the man bit the dog.

A string of UFO sightings in Pueblo, Colorado, hit the front pages of newspapers across the United States. Starting on the night of May 5th, for six nights, the citizens of Pueblo, including the Ground Observer Corps, saw UFO's zip over their community. As usual there were various descriptions but everyone agreed "they'd never seen anything like it before."

On the sixth night, the Air Force sent in an investigator and he saw them. Between the hours of 9:00P.M. and midnight he saw six groups of triangular shaped objects that glowed "with a dull fluorescence, faint but bright enough to see." They passed from horizon to horizon in six seconds.

The next day this investigator was called back to Colorado Springs, his base, and a fresh team was sent to Pueblo.

The man *really* chomped down on the dog in July and the UFO *really* made headlines.

Maybe it was because a fellow newspaper editor was involved, along with the Kansas Highway Patrol, the Navy and the Air Force. Or, maybe it was simply because it was a good UFO sighting.

About the time Miss Iowa was being judged Miss USA in the 1956 Miss Universe Pageant at Long Beach, the city editor of *Arkansas City Daily Traveler*, and a trooper of the Kansas State Highway Patrol were sitting in a patrol cruiser in Arkansas City. It was a hot and muggy night. Occasionally the radio in the cruiser would come to life. An accident near Salina. A drunk driving south from Topeka. Another accident near Wichita. But generally South Central Kansas was dead. The newspaper editor was about ready to go home—it was 10 o'clock—when the small talk he and the trooper had been making was brought to an abrupt finale by three high pitched beeps from the cruiser's radio. An important "all cars bulletin" was coming. Twenty- five years as a newspaperman had trained the editor to always be on the alert for a story so he reached down and turned up the volume. Within seconds he had his story.

"The Hutchinson Naval Air Station is picking up an unidentified target on their radar," the voice of the dispatcher said, with as much of an excited tone as a police dispatcher can have. "Take a look."

Then the dispatcher went on to say that the target was moving in a semi-circular area that reached out from 50 to 75 miles east of Hutchinson. A B-47 from McConnell AFB at Wichita was in the area, searching. The last fix on the object showed it to be near Emporia, in Marion County.

The two men in the patrol cruiser looked at each other for a second or two. Like all newspaper editors, this man had had his bellyful of flying saucer reports—but this was a little different.

"Let's go out and look," he said, fully doubting that they would see anything.

They drove to a hill in the north part of the city where they could get a good view of the sky and parked. In a few minutes an Arkansas City police car joined them.

It was a clear night except for a few wispy clouds scattered across the north sky.

They waited, they looked and they saw.

Shortly before midnight, off to the north, appeared "a brilliantly lighted, teardrop shaped, blob of light." "Prongs, or streams of bright light, sprayed downward from the blob toward the earth." It was big, about the size of a 200 watt light bulb.

As the group of men silently watched, the weird light continued to drift and for many minutes it moved vertically and horizontally over a wide area of the sky. Then it faded away.

As one of the men later told me, "I was glad to see it go; I was pooped."

The next morning literally hundreds of people spent hours conjecturing and describing. After all these years of talk they'd actually seen one. Several photos, showing the big blob of light, were shown around, and two fishermen readily admitted they'd packed up their poles and tackle boxes and headed home when they saw it.

Editor Coyne summed up the feeling of hundreds of Kansans when he said: "I have tended to discount the stories about flying objects, but, brother, I am now a believer."

What was it? First of all it was confusion. Early the next morning Air Force investigators flooded the area asking *the* questions: "What size was it in comparison to a key or a dime?" "Would it compare in size to a light bulb?" "Was there any noise?"

As soon as they left, the military tersely announced that no radar had picked up any target and no B-47's had been sent out. Then they pulled the plugs on the incoming phone lines. The confusion mounted when newsmen tapped their private sources and learned that a B-47 *had* been sent into the area.

A few days later the Air Force told the Kansans what they'd seen: The reflection from burning waste gas torches in a local oil field.

This was greeted with the Kansan version of the Bronx Cheer.

Nineteen hundred fifty-six was a big year for Project Blue Book. According to an old friend, Captain George Gregory, who was then Chief of Blue Book, they received 778 reports. And through a lot of sleepless nights they were able to "solve" 97.8% of them. Only 17 remained "unknowns."

Digging through the reports for 1956, outside of the ones already mentioned, there were few real good ones.

In Banning, California, Ground Observer Corps spotters watched a "balloon-like object make three rectangular circuits around the town." In Plymouth, New Hampshire, two GOC spotters reported "a bright yellow object which left a trail, similar to a jet, moving slowly at a very high altitude." At Rosebury, Oregon, State Police received many reports of "funny green and red lights" moving slowly around a television transmitter tower. And in Hartford, Connecticut, two amateur astronomers, looking at Saturn through a 4-inch telescope, were distracted by a bright light. Turning their telescope on it they observed a "large, whitish yellow light, shaped like a ten gallon hat." Many other people evidently saw the same UFO because the local newspaper said, "reports have been pouring in."

In Miami, a Pan American Airlines radar operator tracked a UFO at speeds up to 4000 miles an hour. Five of his skeptical fellow radar operators watched and were confirmed.

At Moneymore, Northern Ireland, a "level-headed and God fearing" citizen and his wife captured an 18-inch saucer by putting a headlock on it. They started to the local police station, but put the saucer down to climb over a hedge, and it went whirling off to the hinterlands of space.

The 27th Air Defense Division that guards the vast aircraft and missile centers of Southern California was alerted on the night of September 9. In rapid succession, a Western Airlines pilot making an approach to Los Angeles International Airport, the Ground Observer Corps, and numerous Los Angeles citizens called in a white light moving slowly across the Los Angeles basin. When the big defense radars on San Clemente Island picked up an unknown target in the same area that the light was being reported two F-89 jet interceptors were scrambled but saw nothing.

A few days later investigators learned that a $27.65 weather balloon had caused the many thousand dollars' worth of excitement.

The matter of scrambling interceptors has been a sore point with the UFO business for a long time. Many people believe that the mere fact the Air Force will send up two, three, or even four aircraft that cost $2000 an hour to fly is proof positive that the Air Force doesn't believe its own story that UFO's don't exist.

221

The official answer you'll get, if you ask the Air Force, is that they scramble against *any* unknown target as a matter of defense. But over coffee you get a different answer. They write the UFO scrambles off as training cost. Each pilot has to get so much flying time and simulating intercepts against an unidentified light is more interesting than merely "burning holes in the air."

If appropriations are ever cut to the point where training must be curtailed, and Heaven forbid, there will be no more scrambles after flying saucers.

And the colonel who told me this was emphatic.

The year 1957 was heralded in by a startling announcement which ended a long dry spell of UFO news.

At a press conference in Washington, D.C., Retired Admiral Delmer S. Fahrney made a statement. Newspapers across the country carried it complete, or in part, and people read the statement with interest because Admiral Fahrney is well known as a sensible and knowledgeable man. He had fought for and built up the Navy's guided missile program back in the days when people who talked of ballistic missiles and satellites *had* to fight for their beliefs.

First, Admiral Fahrney announced that a non-profit organization, the National Investigations Committee On Aerial Phenomena (NICAP) had been established to investigate UFO reports. He would be chairman of the board of governors and his board would consist of such potent names as:

Retired Vice Admiral R. H. Hillenkoetter, for two years the director of the supersecret Central Intelligence Agency.

Retired Lieutenant General P. A. del Valle, ex-commanding general of the famous First Marine Division.

Retired Rear Admiral Herbert B. Knowles, noted submariner of World War II.

Then Admiral Fahrney read a statement regarding the policies of NICAP. It was as follows:

"Reliable reports indicate that there are objects coming into our atmosphere at very high speeds . . . No agency in this country or Russia is able to duplicate at this time the speeds and accelerations which radars and observers indicate these flying objects are able to achieve.

"There are signs that an intelligence directs these objects because of the way they fly. The way they change position in formations would indicate that their motion is directed. The Air Force is collecting factual data on which to base an opinion, but time is required to sift and correlate the material.

"As long as such unidentified objects continue to navigate through the earth's atmosphere, there is an urgent need to know the facts. Many observers have ceased to report their findings to the Air Force because of the seeming frustration—that is, all information going in, and none coming out. It is in this area that NICAP may find its greatest mission.

"We are in a position to screen independently all UFO information coming in from our filter groups.

"General Albert C. Wedemeyer will serve the Committee as Evaluations Adviser and complete analyses will be arranged through leading scientists. After careful evaluation, we shall release our findings to the public."

Donald Keyhoe, a retired Marine Corps Major, and author of three top seller UFO books, was appointed director. The mere fact that another civilian UFO investigative group was being born was neither news nor UFO history because since 1947 well over a hundred such organizations had been formed. Many still exist; many flopped. But none deserve the niche in UFO history that does NICAP. NICAP had power and it raised a storm that took months to calm down.

NICAP got off to a fast start. Dues were pegged at $7.50 a year, which included a subscription to the very interesting magazine *The UFO Investigator*, and the operation went into high gear.

With such names as Fahrney, Wedemeyer, Hillenkoetter, Del Valle and Knowles for prestige, and Keyhoe for intrigue, saucer fans all over the United States packaged up their seven-fifty and mailed it to headquarters. Each, in turn, became a "listening post" and an "investigator."

Keyhoe set up a Panel of Special Advisors, all saucer fans, to "impartially evaluate" the UFO reports ferreted out by the "listening posts," based on facts uncovered by the "investigators."

Even though the "leading scientists" Fahrney mentioned in his statement never materialized NICAP was cocked, primed, and ready.

To get things off to a gala start Keyhoe, as director of NICAP, wrote to the Air Force and set out NICAP's Eight Point Plan. In essence this plan suggested (some say demanded) that the Air Force let NICAP ride herd on Project Blue Book.

First of all, NICAP wanted its Panel of Special Advisors to review and concur with all of the conclusions on the thousands of UFO reports that the Air Force had in its files.

This went over like a worm in the punch bowl.

First of all, the Air Force didn't feel it was necessary to review its files. Secondly, they knew NICAP. If every balloon, planet, airplane, and bird that caused a UFO report hadn't been captured and a signed confession wrung out, the UFO would be a visitor from outer space.

The Air Force decided to ignore NICAP.

But NICAP wouldn't be ignored. They bombarded everyone from the Secretary of the Air Force on down with telephone calls, telegrams and letters.

Still the Air Force remained silent.

Then NICAP headquarters called in the troops and members from all corners of the nation cut loose. The barrage of mail broke the log jam and just enough information to constitute an answer dribbled out of the Office of the Secretary of the Air Force.

But this didn't satisfy Keyhoe or his UFO hungry NICAPions. They wanted blood and that blood had to taste like spaceships or they wouldn't be happy. The cudgel they picked up next was powerful.

The Air Force had said that there was nothing classified about Project Blue Book yet NICAP hadn't seen every blessed scrap of paper in the Air Force UFO files. This was unwarranted censorship!

While Congress was right in the middle of such important and crucial problems as foreign policy, atomic disarmament, racketeering, integration and a dozen and one other problems, NICAP began to bedevil every senator and representative who was polite enough to listen.

It's the squeaky wheel that gets the grease and in November 1957, the United States Senate Committee on Government Operations began an inquiry concerning UFO's.

I gave my testimony and so did others who had been associated with Project Blue Book.

A few weeks later the inquiry was dropped.

But NICAP had made its name. Of all of the thorns that have been pounded into the UFO side of the Air Force, NICAP drove theirs the deepest.

In the midst of all this mess Admiral Fahrney, General Wedemeyer and General del Valle, politely, and quietly, resigned from NICAP's board of governors.

Neither the loss of these famous names nor the defeat at the hands of the Air Force has stopped NICAP. They continue to forge ahead, undaunted.

In many UFO incidents they have actually uncovered additional, and sometimes interesting, information.

NICAP Director Don Keyhoe has taken a beating, being accused of profiteering, trying to make headlines, and other minor social crimes. But personally I doubt this. Keyhoe is simply convinced that UFO's are from outer space and he's a dedicated man.

While the big NICAP-Air Force battle was going on the UFO's were not waiting to see who won. They were still flying.

At Ellington AFB, Texas, a Ground Observer Corps team spotted a UFO and passed it on to a radar crew. Although the radar crew couldn't pick it up on their sets they saw it visually. The lieutenant in charge told investigators how it crossed from horizon to horizon in 45 seconds.

On March 9, several passengers on a New York to San Juan, Porto Rico airliner were injured when the pilot pulled the big DC-6 up sharply to miss a "large, greenish white, clearly circular-shaped object" which was on a collision course with the plane. The pilots of several other airliners in the same airway confirmed the sighting.

Two weeks later jet interceptors were scrambled over Los Angeles to look for a UFO.

According to the records, the first report of the brilliant and mysterious, flashing, red light came from a man in the east part of Pasadena. But his report was quickly lost in the shuffle as more and more calls began to come in. As the flashing light crossed the Los Angeles Basin from southeast to northwest hundreds of people saw it. Traffic was tied up on the Rose Parade famous Colorado Boulevard as drivers stopped their cars to get out and look.

As it neared the Air Defense Command Filter Center in Pasadena the filter center personnel, those that could be spared, went out and looked. They saw it. Police switchboards lit up a solid red as it crossed the San Gabriel Valley.

Near midnight a CAA radar picked up unidentified targets near the Oxnard AFB, at Oxnard, California (northwest of Los Angeles), and at almost that identical time people on the airbase saw the light

This did it, and two powerful jets, equipped with all weather radar, came screaming into the area.

But it was the same old story—no contact—the UFO was gone.

The midwest was visited on the morning of May 23rd, when five observers in Kansas City saw four silver, disc-shaped objects flying in formation at extremely high speed. At one point during their flight two of the objects broke formation and veered off but soon rejoined. It took the objects only four minutes to cross the sky.

There were other reports during the first half of 1957, 250 of them to be exact, and many could be classified as "good." But they were nothing compared to those that were to come.

On November 3, 1957, a rash of sightings broke out in Texas and they had a brand new twist. To do things up right the powers that guide the UFO picked the town of Levelland only 27 miles west of Lubbock, the home of the now traditional "Lubbock Lights."

It was with a tug of nostalgia that I read about these reports because five years before, almost to the day, Lubbock had plunged the Air Force, and me, into the UFO mystery on a grand scale.

According to the best interpretation of the maze of conflicting stories, facts and rumors about these famous sightings the only positive fact is that there were scattered storm clouds across West Texas on the night of November 4, 1957. This was unusual for November and everyone in the community was just a little edgy.

It was early in the evening, at least early for West Texas on a Saturday night, when Pedro Saucedo, a farm worker, and his friend Joe Salaz, started out in Saucedo's truck toward Pettit, ten miles northwest of Level-land. They had just turned off State Highway 116 and were heading north on a country road when the two men saw a flash of light in an adjacent field. Saucedo, a Korean War Veteran, and Salaz didn't pay much attention to the light at first. They only noticed that it was coming closer. "It seemed to be paralleling us and edging a little closer all the time," Saucedo later recalled. Still neither man paid any attention to the light. They drove on, Saucedo watching the road and Salaz talking.

Then it hit.

The first signal of something wrong was when the truck's headlights went out; then the engine stopped. Before Saucedo could hit the starter again he glanced over his left shoulder. A huge ball of fire was "rapidly drifting" toward the truck. Without a second's hesitation Saucedo did what the Korean

War had taught him to do when in doubt, he shoved open the car door and hit the dirt.

Salaz just sat.

"The 'Thing' passed directly over my truck with a great sound and rush of wind," Saucedo later told County Sheriff Weir Clem, after he'd started his truck and had driven back to Levelland. "It sounded like thunder and my truck rocked from the blast. I felt a lot of heat."

The "Thing," which disappeared across the prairie, looked like a "fiery tornado."

Five years before and a little east of where Saucedo and Salaz were "buzzed" I had talked to two women who described almost an identical UFO. And it remains "unknown" to this day.

In Levelland, the two men's story would have been enough to keep Sheriff Clem busy for the rest of the night but between the hours of 8:15P.M. and midnight on the 2nd the "Levelland Thing" struck five more times.

James D. Long, a Waco truck driver, came upon "it" four miles west of Levelland and fainted as it roared over his truck. Ronald Martin, another truck driver, was stopped east of Levelland, as was Newell Wright, a Texas Tech student. Jim Wheeler, Jose Alvarez and Frank Williams added their stories to the melee.

All of those who had been attacked told Sheriff Clem a similar story: "The 'Thing' was shaped something like an egg standing on end. It was fiery red, more like a red neon light. It was about 200 feet long and was about 200 feet in the air. When it came close to cars the engines would stop and the lights would go out."

"Everyone," Sheriff Clem said, "seemed very excited."

That night everyone in West Texas saw UFO's. Sheriff Clem saw a brilliant light in the distance. Highway patrolmen Lee Hargrove and Floyd Cavin reported similar brilliant lights at the same time but from a different location. The control tower operators at the Amarillo Airport, to the north, saw a "blue, gaseous object which moved swiftly and left an amber trail."

There were dozens more. It was a memorable Saturday night in Levelland.

But unbeknown to Sheriff Clem or the residents of West Texas, they weren't alone on the visitor's list.

At 2:30A.M. on Sunday morning, only a few hours after the "Thing" raised havoc around Levelland, an army military police patrol was cruising the supersecret White Sands Proving Ground in New Mexico.

Here is their report as they gave it to Air Force UFO investigators:

"At approximately 0230, 3 November 1957, Source, together with PFC —
——, were on a routine patrol of the up range area of the White Sands Proving Ground when Source noticed a 'very bright' object high in the sky. This object slowly descended to an altitude estimated to be approximately 50 yards where it remained motionless for about 3 minutes, then it descended to the ground where the light went out. The object was not blurred or fuzzy,

emitted no vapor or smoke. The object was in view for about 10 minutes, and Source estimated that it was approximately 2 or 3 miles away. It was estimated to be between 75 and 100 yards in diameter and shaped like an egg. Source stated that it was as large as a grapefruit held at arm's length. The weather was cold, drizzling and windy, and Source stated no stars were visible. After the light went out Source and PFC ——— continued north to the STALLION SITE CAMP and reported the incident to the Sergeant of the Guard who returned to the area but failed to find anything."

The flap was on.

On Monday, the 4th, the "Levelland Thing" struck again near the White Sands Proving Ground. James Stokes, a 20-year Navy veteran, and an electronics engineer, had the engine of his new Mercury stopped as "a brilliant, egg-shaped" object made a pass at the highway. As it went over, Stokes said, "it felt like the radiation of a giant sun lamp."

Stokes said there were ten other carloads of people stopped but if this is true no one ever found out who they were.

The Air Force wrote off Stokes' story as, "Hoax, presumably suggested by the Levelland, Texas, reports."

Maybe the Air Force didn't believe James Stokes but when the Coast Guard Cutter *Seabago* radioed in their report from the Gulf of Mexico wheels began to turn—fast.

On Tuesday morning, the 5th, the *Seabago* was about 200 miles south of the mouth of the Mississippi River on a northerly heading. At 5:10A.M. her radar picked up a target off to the left at a distance of about 14 miles. This was really nothing unusual because they were under heavily traveled air lanes.

The early morning watch is always rough and as the small group of officers and men in the Combat Information Center quietly watched the target, with a noticeable lack of enthusiasm, it moved south, made a turn, and headed back to the north again. A few of the men noticed that the turn looked "a little different," but this early in the morning they didn't give it much thought.

At 5:14 the target went off the scope to the north.

At 5:16 it was back and the lassitude was instantly gone. Now the target was 22 miles *south* of the ship. No one in the CIC had to draw a picture. Something, in two minutes, had disappeared off the scope to the north, made a big swing around the ship, out of radar range, and had swung in from the south!

Word went up to the lookouts. They tensed up and began to scan the sky.

The radar contacts continued.

This second contact, south of the ship, was held for two full minutes as the target moved out from 22 to 55 miles. Then it faded.

At 5:20 the target was back but now it was *north* of the ship again, and it was hovering!

Again the lookouts were called. Could they see anything now? Their "No" answers didn't hold for long because seconds later their terse reports began to come into the CIC. A "brilliant light, like a planet" was streaking across the northwest sky about 30 degrees above the horizon. Unfortunately the radar had lost contact for a moment when the visual report came in.

At 5:37 the target disappeared from the scopes and was gone for good.

The *Seabago* Case was ended but the UFO's continued to fly.

Reports continued to come into the Air Force and a lot of investigators lost a lot of sleep.

The next day at 3:50P.M. the C.O. of an Air Force weather detachment at Long Beach, California, and twelve airmen watched six saucer- shaped UFO's streak along *under* the bases of a 7000 foot high cloud deck.

On the same day, also in Long Beach, officers and men at the Los Alamitos Naval Air Station saw UFO's almost continuously between the hours of 6:05 and 7:25P.M.

Long Beach police reported "well over a hundred calls" during this same period.

During November and December of 1957 it was a situation of you name the city and there was a UFO report from there. Trying to sift them out and put them in a book would be like sorting out a plateful of spaghetti. And if you succeeded you would have a document the size of the New York City telephone directory.

Most of the reports were explained.

The Levelland, Texas, sightings were written off as "St. Elmo's Fire." The military police at the White Sands Proving Ground saw the moon through broken clouds and the crew of the Coast Guard ship *Seabago* were actually tracking several separate aircraft.

The 1957 flap was as great as the previous record breaking 1952 flap. During 1957 the Air Force received 1178 UFO reports. Of these, only 20 were placed on the "unknown" list.

In comparison to 1957, the first months of 1958 were a doldrums. Reports drifted in at a leisurely pace and the Air Force UFO investigating teams, blooded during the avalanche of 1957, picked off solutions like knocking off clay pipes in a shooting gallery.

In Los Angeles, a few clear nights drove the Air Defense Command nuts. People could actually see the sky and the sight of so many stars frightened them.

Unusual atmospherics in Georgia made stars jump and radars go crazy; and a balloon, hanging over Chicago at dusk, cost the taxpayers another several thousand dollars but the pilots made their flight pay.

A statement by Dr. Carl Jung, renowned Swiss psychologist, was widely publicized in July 1958. Dr. Jung was quoted as saying, in a letter to a U.S. saucer club, "UFO's are real." When Dr. Jung read what he was supposed to have written the Alps rang with screams of "misquote."

No one got excited until the early morning of September 29th.

Shortly before dawn on that day a confusing mess of reports began to pour into the Air Force. Some came from the Washington, D.C., area. People right in NICAP's backyard told of seeing a "large, round, fiery object" shoot across the sky from southeast to northwest. A few excited observers, all from the country northwest of Washington, "had seen it land" and even as they telephoned in their reports they could see it glowing behind a neighbor's barn.

Other reports, also of a "huge, round, fiery object," came in from such places as Pittsburgh, Somerset, and Bedford, all in Pennsylvania; and Hagerstown and Frederick in Maryland. To add to the confusion, people in Pennsylvania reported seeing three objects "flying in formation."

When the dust settled Air Force investigators took the first step in the solution of any UFO report. They plotted the sightings on a map, and collated the directions of flight, descriptions and times of observation. It was obvious that the object had moved along a line between Washington, D.C., and Pittsburgh. It was traveling about 7000 miles an hour and everyone had obviously seen the same object. By the time it had passed into Pennsylvania it had split into three objects.

But the hooker was the reported landings northeast of Washington. Too many people had reported a glow on the ground to write this factor off even though an investigator, dispatched to the scene shortly after dawn, had found nothing in the way of evidence.

One possibility was that some unknown object had streaked across the sky, landed and then took off again.

Could be, but it wasn't.

The next night the case broke. The glow from the landing was a bright floodlight on a barn. No one had ever really noticed it before until the object passed nearby.

A few days later the object itself was identified. From the many identical descriptions Project Blue Book's astrophysicist pinned it down as a large meteor. The meteor had broken up near the end of its flight to produce the illusion of three objects flying in formation.

Of all the 590 UFO reports the Air Force received in 1958, probably the weirdest was solved before it was ever reported.

About four o'clock on the afternoon of October 2, 1958, three men were standing in a group, talking, outside a tungsten mill at Danby, California, right in the heart of the Mojave Desert The men had been talking for about five minutes when one of them, who happened to be facing the northwest, stopped right in the middle of a sentence and pointed. The other two men looked and to their astonishment saw a brilliant glow of light. It was so close to the horizon that it was difficult to tell if it was on the horizon or in the air just above it.

At first the men ignored the light but as it persisted they became more interested. They'd all heard "flying saucer" stories and, they later admitted, this possibility entered their minds.

As they watched they speculated. It could be something natural but all of them had been around this area for months and they'd never seen this light before. About the time they decided to get a telescope and take a closer look the light suddenly faded.

All the next day the men kept glancing off toward the northwest as they worked but the clear blue sky was blank. Then, at 4:00P.M., the light was back. This time they had a telescope.

All the men took turns looking at the object and all agreed that it was about 15 feet long, 5 feet high and solid. It looked like the sun reflecting off shiny metal. It was about four miles away, they estimated, and almost exactly on the horizon.

Now the men's curiosity was thoroughly whetted. Martian spaceship or whatever, they were going after it. But a several-hour search of the area produced nothing. And, as soon as they left the mill they lost sight of the object.

Darkness brought the search to a halt.

The next day at 4:00P.M. a crowd had gathered and the UFO kept its appointment. Again the men studied the object and tension ran high.

Someone had resurrected the stories of UFO's landing in the desert. At the time they'd sounded absurd but now, standing there looking at a UFO, it was different.

A party of men were all ready to jeep out into the desert to make another search when one of them made a discovery. There were guy wires coming out of the UFO and running down into the trees. Other people looked. And then the solution hit like a fireball.

Exactly in line with the UFO, and ten miles away, not four, was a set of antennas for the California State Highway Patrol radio. The sun's rays were reflecting from these antennas. They'd never seen this before because on only a few days during the year was the sun at exactly the right angle to produce the reflection.

The men were right. In a few days the Danby UFO left and it never came back.

Nineteen hundred fifty-eight was not a record year for UFO's. The 590 reports received didn't stack up to the 1178 for 1957, or the 778 for 1956, or the 918 for 1952. But a new record was set when the percentage of unknowns was pared down to a new low. During 1958 only 9/10 of one per cent of the reports, or 5 reports, were classified as "unknown."

More manpower, better techniques, and just plain old experience has allowed the Air Force to continually lower the percentage of "unknowns" from 20%, while I was in charge of Project Blue Book, to less than 1%, today.

No story of the UFO would be complete without describing one of these unknowns, so here's one exactly as it came out of the Project Blue Book files:

"On 31 October 1958, this Center received a TWX reporting an UFO near Lock Raven Dam. A request for a detailed investigation was sent to the nearest Air Force Base. The following is a summary of the incident and subsequent investigation:

"Two civilians were driving around near Lock Raven Dam on the evening of 26 October 1958. When they rounded a curve about 200 to 300 yards from a bridge they saw what appeared to be a large, flat, egg shaped object hovering about 100 to 150 feet above the bridge superstructure. They slowed their car and when they got to within 75 or 80 feet of the bridge their engine quit and their lights went out. The driver immediately stepped on the brakes and stopped the car. Attempts were made to start the car and when this was unsuccessful they became frightened and got out of the car. They put the car between them and the object and watched for approximately 30 to 45 seconds. The object then seemed to flash a brilliant white light and both men felt heat on their faces. Then there was heard a loud noise and the object began rising vertically. The object became very bright while rising and its shape could not be seen as it rose. It disappeared in five to ten seconds.

"After the object disappeared, the car was started and they turned it around and drove to where a phone was located and contacted the Towson Police Department. Two patrolmen were sent to meet them. The two men told the patrolmen of their experience. The witnesses then noticed a burning sensation on their faces and became concerned about possible radiation burns. They went to a Baltimore Hospital for an examination. Both witnesses were advised by the doctor that they had no reason for concern.

"An extensive investigation was made concerning this incident. However, no valid conclusion could be made as to the possible nature of the sighting and it remains unidentified."

So ended 1958 and in its final tally of sightings for the year Project Blue Book added a new space age touch—earth satellites had accounted for eleven UFO reports.

Nineteen hundred fifty-nine came in with a good one. We used to call these reports "Ground-air-visual-radar" sightings and they make interesting reading.

At Duluth, Minnesota, in March, it's dark by five o'clock in the evening. It's cold. The temperature hovers around zero and it's so clear you have a feeling you can almost reach up and touch the stars.

It was this kind of a night on March 13, 1959, and as the officers and men of the Air Defense Command fighter squadron at the Duluth Municipal Airport moved, they shuffled along slowly because the heavy parkas and arctic clothing they wore were heavy.

Then came the UFO report and things speeded up.

At 5:20P.M., exactly, the operations officer noted the time, word came in over the comm line that someone had sighted an unidentified flying object off to the north. Word flashed around the squadron and as people rushed out of buildings to look they were joined by those already outside.

And there it was: big, round and bright, and it was moving at high speed. Some observers thought it was "greenish," others "reddish," but it was something and it was there.

The bearing was 300 degrees from the base.

It was an awesome sight and it became even more awesome when a quick call to an adjacent radar site brought back the word that they had just picked up a target on a bearing of 300 degrees from the air base. They were tracking it and taking scope photos.

In the alert hangar, the two pilots standing the alert had been listening to a running account of the sighting so when the scramble bell rang they took off for their airplanes like a couple of sprinters.

As the two big alert hangar doors swung up the whining screech of the jet starters, followed by thunder of the engines, filled the airfield. The atmosphere around the Duluth Municipal Airport was closely akin to Santa Anita the instant the starting gates open.

I've been around when jet interceptors scramble and you can twang the tension with your finger.

As the people on the ground watched they could first see the flame of the jet's afterburner disappear into the night. Then the jet's navigation lights faded out on a bearing of 300 degrees.

At the radar site they still had the target and there were many excited people watching the big pale, orange scopes as two little bright points of light began to close on a bigger blob of light.

Then the pilots gave the "Tally-ho"—they were in visual contact.

But the "Tally-ho" had no more been given than the big blob of light on the target began to pull away from the fighters and was soon off the scope.

The pilots kept visual contact, though, and the radio provided the details of the chase to the now blind crew in the radar room.

The two jets bored north, with afterburner on, and the needles on their machmeters passed the "1.0" mark. But still the UFO was just as far away as it had ever been.

The chase went on for a few minutes more before the pilots pulled their throttles back into the cruise position, turned, and came home.

Even before they landed, the people at the airbase saw the big, round and bright UFO rapidly begin to fade and then it was gone.

So ended the glamour and the dog work began.

Each man who had seen the UFO visually was carefully interrogated. Weather reports were collected. Radarscope photos were developed. The two pilots received special attention. The exact bearing of the UFO was measured and 300 degrees magnetic was correct.

The bundle of data was packed up and sent to Project Blue Book. The panel of experts convened.

First, the radarscope photos were examined.

"Those targets could be interference from other radars," said the radar expert, and he mentally ticked off a dozen and one other similar cases of known interference. The weather data, and locations and frequencies of other radars were checked out.

Beyond doubt it was interference from another radar that caused the target.

Now, the visual sighting.

Balloon? No, the fighters could have caught a balloon in seconds.

Airplane? Same answer. These jets were the fastest things in the air.

Planet or star? Out came the almanacs and the puzzle went to the astrophysicist. Venus was on a bearing of 300 degrees from the Duluth Municipal Airport at 5:20P.M. on March 23rd. *But* Venus was just below the horizon at that time and the observers said the UFO was "moving fast."

Once again the weather charts were studied. The atmospheric conditions were such that it was very possible that due to refraction Venus would have been visible just on the horizon. The fact that the UFO faded so fast would bear this out because the conditions for such refraction are critical and a slight change in atmospheric conditions could easily have caused the planet to disappear.

The speed—a common illusion. Further interrogation of the observers showed it had never moved.

So, the history of the UFO is almost brought up to date.

Chapter Nineteen - Off They Go into the Wild Blue Yonder

At 12:30P.M. on Thursday, November 20, 1952, history was made.

At least, so says George Adamski, lecturer on philosophy and student of technical matters and astronomy.

At 12:30P.M. on Thursday, November 20, 1952, George Adamski was the first man on earth to talk to a Venusian.

At least, so says George Adamski.

I was chief of Project Blue Book at the time and the name "Professor Adamski"—he had a title then—wasn't new to me. He, or some of his followers had been showering the Air Force with photos of flying saucers. Letters by the gross were coming in demanding recognition of the great professor and an analysis of his photos.

We obliged and the photos were examined by the experts at Wright- Patterson Photo Reconnaissance Labs. The verdict came back: "They could be genuine, of course, but they also could have been easily faked by a ten year old with a Brownie camera."

For a few weeks we forgot George Adamski. But then the press began to clamor at our gates. The news was leaking out of Southern California. George Adamski had talked to a Venusian! We held out for a long time but the pressure mounted and I headed for California to find out what it was all about.

As far as George Adamski was concerned I was just another thirsty sightseer from the famous observatory on Mt. Palomar when I walked into the little restaurant at the foot of this famous mountain one day in 1953.

The four stool restaurant, with a few tables, where Adamski worked as a handyman, was crowded when I arrived and he was circulating around serving beer and picking up empty bottles. There was no doubt as to who he was because his fame had spread. To the dozen almost reverently spoken queries, "Are you Adamski?" he modestly nodded his head.

Small questions about the flying saucer photos for sale from convenient racks led to more questions and before long the good "professor" had taken a position in the middle of the room and was off and running.

In his slightly broken English he told how he was the son of poor, Polish immigrants with hardly any formal education.

To look at the man and to listen to his story you had an immediate urge to believe him. Maybe it was his appearance. He was dressed in well worn, but neat, overalls. He had slightly graying hair and the most honest pair of eyes I've ever seen.

Or maybe it was the way he told his story. He spoke softly and naively, almost pathetically, giving the impression that "most people think I'm crazy, but honestly, I'm really not."

Adamski started his story by telling how he had spent many long and cold nights at his telescope "at the request of the government" trying to photograph one of the flying saucers everyone had been talking about. He'd been successful, as the full photograph racks on the wall showed, and he thought the next step would be to actually try to contact a saucer.

For some reason, Adamski didn't know exactly why, on November 19th he'd decided to go out into the Mojave Desert. He'd called some friends and told them to meet him there.

By noon the next day the party, which consisted of Adamski and six others, had met and were eating lunch near the town of Desert Center on the California-Arizona border.

They looked for saucers, but except for an occasional airplane, the cloudless blue sky was empty. They were about ready to give it up as a bad day when another airplane came over. Again they looked up, but this time, in addition to seeing the airplane, they saw a silvery, cigar-shaped "flying saucer."

For some reason, again he didn't know why, the group of people moved down the road where Adamski left them and took off into the desert alone.

By this time the "space ship" had disappeared and once again Adamski was about to give up.

Then, a flash of light caught his eye and a smaller saucer (he later learned it was a "scout ship") came drifting down and landed about a half mile from him. He swung his camera into action and started to take pictures. Unfortunately, the one picture Adamski had to show was so out of focus the scout ship looked like a desert rock.

He took a few more pictures, he told his audience, and had stopped to admire the little scout ship when he suddenly noticed a man standing nearby.

Now, even those in the crowded restaurant who had been smirking when he started his story had put down their beers and were listening. This is what they had come to hear.

You could actually have heard the proverbial pin drop.

Adamski told what went through his mind when he first saw the man—maybe a prospector. But he noticed the man's long, shoulder-length, sandy-colored hair, his dark skin, his Oriental features and his ski- pant type trousers. He was puzzled.

Then it came into his mind like a flash, he was looking at a person from some other world!

Through mental pictures, sign language, and a few words of English, Adamski found out the man was from Venus, he was friendly, and that they (the Venusians) were worried about radiation from our atomic bombs.

They talked. George pointed to his camera but the man from Venus politely refused to be photographed. Adamski pleaded to go into the "ship" to see how it operated but the Venusian refused this, too.

They talked some more—of spaceships and of solar systems—before Adamski walked with his new found friend to the saucer and saw the Venusian off into space.

At this point Adamski recalled how he had glanced up in the sky to see the air full of military aircraft.

Needless to say, the rest of Adamski's party, who had supposedly seen the "contact" from a mile away, were excited. They rushed up to him and it was then that they noticed the footprints.

Plainly imprinted in the desert sand were curious markings made by ridges on the soles of the Venusian's shoes.

At the urging of the crowd in the restaurant Adamski took an old shoe box out from under the counter. One of his party, that day, had just happened to have some plaster of paris and the shoe box contained plaster casts of shoe prints with strange, hieroglyphic- like symbols on the soles. No one in the restaurant asked how the weight of a mere man could make such sharp imprints in the dry, coarse desert sand.

Next he showed the sworn statements of the witnesses and the crowd moved in around him for a better look.

As I left he was graciously filling people in on more details and the cash register was merrily ringing up saucer picture sales.

I didn't write the trip off as a complete loss, the weather in California was beautiful.

Adamski held the UFO spotlight for some time.

The Venusians paid him another visit, this time at the restaurant, and he photographed their "ship." This, whether by Venusian fate or design, increased the flow of traffic to the restaurant at the base of Mt. Palomar.

It also had its side effects.

An astronomer from the observatory that houses the world famous 200-inch telescope on top of Mt. Palomar told me: "I hate to admit it but the number of week end visitors has picked up here. People drive down to hear George and decide that since they're down here they might as well come up and see our establishment."

But George Adamski didn't hold the front center of the stage for long. In rapid succession others stepped forward and hesitantly admitted that they too had been contacted.

Truman Bethurum, a journeyman mechanic of Redondo Beach, California, was next up.

Actually, he admitted, *he* had been the first earthman to talk to a person from another world. Back on the night of July 26, 1952, four months before Adamski, a group of eight or ten, short, olive-skinned men with black wavy hair, had awakened him while he was asleep in a truck in the desert near Mormon Flats, Nevada.

These little men, unlike Adamski's, spoke any language.

"You name it," they'd quipped to Bethurum, "we speak it."

In a newspaper article that was voted "Best Read of 1953," Bethurum told how the little men he met had been more cooperative and had actually taken him into their saucer, a huge job 300 feet in diameter and 16 feet high.

Once inside, Bethurum had met the captain of the "scow"—a true leader of men. Aura Rhanes was her name and she was a Venus de Milo with arms and warm blood. "When she spoke her words rhymed." They chatted and Bethurum learned that he was on the "Admiral's scow" the command ship of Clarion's fleet of saucers.

All in all, Bethurum made eleven visits to Aura's scow. Each time they'd sit and talk. Bethurum told her about the earth and she told of the idyllic, Shangri-La type planet of Clarion—a yet undiscovered planet which is always opposite the moon.

But before too long, both Truman Bethurum and George Adamski had to move over. Daniel Fry, an engineer, stepped in.

At a press conference to kick off the International Saucer Convention in Los Angeles, Fry told how he had not only contacted the spacemen *two years before* Adamski and Bethurum, he had actually *ridden* in a flying saucer.

It had all started on the night of July 4, 1950, when engineer Fry was temporarily employed at White Sands Proving Ground in New Mexico.

It was a hot night, and with nothing else to do, Fry decided to take a walk across the desert. He hadn't traveled far when he saw a bluish light hovering over the mountains which rim this famous proving ground. He paid no attention. He'd heard flying saucer stories before and just plain didn't believe them.

But as he watched, the light came closer and closer and closer, until a weird craft came silently to rest on the desert floor not seventy feet away.

For seconds, Fry, who had seen missile age developments at White Sands that would have dumfounded most laymen, merely stood and stared.

The object, Fry told newsmen, was an "ovate spheroid about thirty feet at the equator." (Fry has a habit of drifting off into the technical). Its outside surface was a highly polished silver with a slight violet iridescent glow.

At first Fry wanted to run but his rigid technical training overrode his common, natural urges. He decided to go over to the object and see what made it tick.

He circled it several times and nothing broke the desert silence. Then he touched it.

"Better not touch that hull, pal, it's hot," boomed a voice in a Hollywoodian tone.

Fry recoiled.

The voice softened and added, "Take it easy, pal, you're among friends."

After politely reading off the spaceman, or whoever he was, for scaring him, pal Fry and the voice settled down for a friendly moonlight chat. Fry learned that the voice was indeed that of a spaceman and they were down to pick up a new supply of air. After about four years of earth air transfusions, according to the spaceman, they would become adapted to our atmosphere, and our gravity, and become "immunized to your bi-otics." The craft, Fry was told, was a "cargo carrier," unmanned and built to zoom down and scoop up earth air.

The conversation went on, waxing technical at times, and ended with an invitation to look into the ship. Then the spaceman, possibly carried away by all the interest Fry was showing, offered a ride.

Fry accepted and they antidemagnetized off for New York City. Thirty minutes later they were back at White Sands.

Over New York City they came down from 35 to 20 miles and Fry could read the marquee of the Fulton Theater. "The Seven Year Itch" was playing.

He hadn't told the Air Force about his ride before because he was afraid he'd lose his job. But, at the press conference, he did plug his new book, *The White Sands Incident.*

By this time Adamski had already published his book *Flying Saucers Have Landed* and it looked as if Fry was going to cut him out. But Fry took a lie detector test on a widely viewed West Coast television show and flunked it flat.

His stock dropped as fast as it had risen but the decline was somewhat checked when a well known Southern California medium wrote to "her old friend" J. Edgar Hoover about the situation. Hoover, the story goes, shot back an answer—lie detectors are no good.

But the damage had been done. The "rigged" lie detector test had unfortunately relegated Daniel Fry, "engineer," "missile expert," "part owner of an engineering plant," and interplanetary hitchhiker to the bush league.

With Adamski and Bethurum on the stage and Fry peeking out of the wings all hell broke loose.

One could say that everyone tried to get into the act, but I'd rather think that each colony of space people tried to promote their own candidate.

In England, one Cedric Allingham met a Martian on the moors. In France, Germany, the United States, Portugal, Brazil, Spain— everywhere—people "too uneducated to pull a hoax" met green men, dark men, white men, big men with little heads, little men with big heads and men with pointed heads. They wore motorcycle belts, baggy pants, diver suits, and were naked.

One lady proudly announced that a Venusian had tried to seduce her and within days another snorted in disgust. A Martian *had* seduced her.

Then Adamski took a hop through outer space and back.

Saucers poured forth words of wisdom via radio, light beams and mental telepathy. All of these messages were duly recorded on tape and sales were hot at $4.50 per 10-minute tape.

Not to be outdone by any other lousy planet, the Venusians picked up a young man from Los Angeles and actually took him to Venus. Not once, but three times.

He packed in audiences by telling how he had been contacted one night and asked by a "strange man" if he would go on an important mission. Afraid, but not one to shirk his patriotic duties, he met the stranger at a prearranged spot and was whisked off to Venus. During a high level conference up there he was given the word: Tell the earthlings to lay off their atomic weapons, or else. They're killing all our doves and we make our flying saucers out of the feathers our live doves shed.

The Venusians, this space traveler warned his audiences, were already infiltrating the earth and he intimated that they were ready to move in case we didn't cease atomic testing.

His next two trips to Venus were purely social.

The highlight of his lecture, when he awes his audience, is when he whips out his proof: (1) a blood smear on a slide—genuine Venusian blood, (2) an affidavit from his landlady stating he wasn't home on three occasions, and (3) a photo of a Venusian walking in Los Angeles' McArthur Park. The mere fact that the Venusian looks like any Joe Doakes walking down the street is a picayunish point. Venusians look just like us.

And it hasn't stopped. During the big UFO flap of 1957 a man stumbled onto a landed saucer and chatted awhile with its occupants. A few months later, soon after the atomic powered *U.S.S. Nautilus* made its historic trip under the polar ice cap, this same man snorted in disgust. He packed his suitcase and started on a lecture tour. Months before *he'd* been there in a flying saucer.

Once again people shelled out hard cash to hear his story.

Wherever you are, Mr. P. T. Barnum, you are undoubtedly grinning from ear to ear.

But there is a sober side to this apparently comical picture. The common undertone to many of these stories "hot from the lips of a spaceman" is Utopia. On these other worlds there is no illness, they've learned how to cure all diseases. There are no wars, they've learned how to live peaceably. There is no poverty, everyone has everything he wants. There is no old age, they've learned the secret of eternal life.

Too many times this subtle pitch can be boiled down to, "Step right up folks and put a donation in the pot. I'm just on the verge of learning the spaceman's secrets and with a little money to carry out my work I'll give *you* the secret."

I've seen a man, crippled by arthritis, hobbling out into the desert in hopes that his "friend who talks to the Martians" could get them to cure him on their next trip. I've seen pensioners, who needed every buck they had, shell out money to "help buy radio equipment" to contact some planet to find out how they'd solved their economic problems. I saw a little old lady in a many times mended dress put down a ten dollar bill to help promote a "peace campaign" backed by the Venusians. She'd lost two sons in the war but had four grandsons she wanted to keep alive. A couple died and left $15,000 to a man to build a "longevity machine" so others could live. The Martians had given him the plans.

A woman died of thirst and exposure in the Mojave Desert trying to reach the spot where a man told her he was going to "make a contact."

Some of it isn't comical.

Even though the field is becoming crowded, through thick and thin, Martian and Venusian, the old Maestro, George Adamski, is still head and shoulders above the rest. The hamburger stand is boarded up and he lives in a big ranch house. He vacations in Mexico and has his own clerical staff. His two books *Flying Saucers Have Landed* and *Inside the Space Ships* have sold something in the order of 200,000 copies and have been translated into nearly every language except Russian. To date, he's had eleven visits from people from Mars, Venus and Saturn. Evidently Truman Bethurum's Aura Rhanes put out the word about earthmen because two beautiful spacewomen have now entered Adamski's life: an "incredibly lovely" blonde named Kalna, and the equally beautiful Illmuth.

Only a few months ago, while on one of his numerous nationwide lecture tours, a saucer unexpectedly picked Adamski up in Kansas City and took him on a galactic cruise before depositing him at Ft. Madison, Iowa, where he had a lecture date. He "wowed" the packed auditorium with his "proof"—an unused Kansas City to Ft. Madison train ticket.

Last week, in the Netherlands (Adamski's nationwide tours have expanded to world-wide tours), he repeated his exploits to Queen Juliana.

But at Buckingham Palace, Mr. Barnum, all he saw was the changing of the guard.

Chapter Twenty - Do They, or Don't They?

During the past four years the most frequent question I've been asked is: "What do you personally think? Do unidentified flying objects exist, or don't they?"

I'm positive they don't.

I was very skeptical when I finished my tour of active duty with the Air Force and left Project Blue Book in 1953, but now I'm convinced.

Since I left the Air Force the Age of the Satellite has arrived and we're in it. Along with this new era came the long range radars, the satellite tracking cameras, and the other instruments that would have picked up any type of "spaceship" coming into our atmosphere.

None of this instrumentation has ever given any indication of any type of unknown vehicle entering the earth's atmosphere.

I checked this with the Department of Defense and I checked this through friends associated with tracking projects. In both cases the results were completely negative.

There's not even a glimmer of hope for the UFO.

Then there's Project MOONWATCH, the Optical Satellite Tracking Program for the International Geophysical Year.

Dr. J. Allen Hynek, the director of MOONWATCH wrote to me: "I can quite safely say that we have no record of ever having received from our MOONWATCH teams any reports of sightings of unidentified objects which had any characteristics different from those of an orbiting satellite, a slow meteor, or of a suspected plane mistaken for a satellite."

Dr. Hynek should know. He has investigated and analyzed more UFO reports than any other scientist in the world.

And the third convincing point is that twelve years have passed since the first UFO report was made and still there is not one shred of material evidence of anything unknown and no photos of anything other than meaningless blobs of light.

The next question that always arises is: "But people are seeing something. Experienced observers, like pilots, scientists and radar operators have reported UFO's."

To be very frank, we heard the words "experienced observer" so many times these words soon began to make us ill.

Everyone, except housewives with myopia, were experienced observers.

Pilots, "scientists" (a term used equally as loosely), engineers, radar operators, everyone who reported a UFO was some kind of an "experienced observer." This man had taught aircraft recognition during World War II. He was an experienced observer. That man spent four years in the Air Force. He was an experienced observer. We soon learned that everyone is an experienced observer as long as what he sees is familiar to him. As soon as he sees something unfamiliar it's a UFO.

Pilots probably come as close to falling into this category as anyone since they do spend a lot of time looking around the sky. But even those who can rattle off the names and locations of stars, planets and constellations don't know about a few relatively rare astronomical phenomena.

The bolide, or super meteor, is a good example. Few pilots have ever, or will ever, see a deluxe model bolide but when they do they'll never forget it. It's like someone shooting a flare in front of your face. There are a number of reports of bolides in the Blue Book files and each pilot who made each report called each bolide a UFO. The descriptions are almost identical to the classic descriptions of bolides found in astronomy books.

While on the subject of meteors, if most people realized that meteors can have a flat trajectory, they can go from horizon to horizon, they can travel in "formation" (groups), and they can be seen in daylight (as "large silver discs"), the work of UFO investigators would be lighter.

Enough of meteors and back to our experienced observers.

The example of pilots and bolides holds true in many, many other cases.

Take high flying jets for example. To a person in an area where there isn't much high altitude air traffic, a thin, blood red streak in the sky at sunset, or shortly after, is a UFO. To anyone in an area where there are a lot of high flying jets even our myopic housewife, it's just another vapor trail. They're as common as the sunset.

When the flashing red strobe lights, now used on practically all aircraft, were still in the experimental stage back in 1951 they gave us fits. Every time an airplane with one of these flashing lights made a flight people within miles, including other pilots, called in UFO reports. Now these strobe lights are common and no one even bothers to look up.

The same held true, and still does, for the odd array of lights used on tanker planes during aerial refueling operations.

Some phenomena are so rare and so little is known about them that they are always UFO's. The most common is the disc following the airplane.

I've never heard an explanation for this phenomenon but it exists and I've seen it on three occasions. Maybe a dense blob of air tears off the airplane, floats along behind, and reflects the sunlight. Whatever it is, it gives the illusion of a saucer "chasing" an airplane. Sometimes it's steady and sometimes it darts back and forth. It only stays in view a few seconds and when it disappears it fades and looks for all the world as if it's suddenly streaking away into the distance.

Birds, bees, bugs, airplanes, planets, stars, balloons, and a host of other common everyday objects become UFO's the instant they are viewed under other than normal situations.

Then there is radar. This poor inanimate piece of electronic equipment has taken a beating when UFO proof is being offered. "Radar is not subject to the frailties of the human mind," is the outcry of every saucer fan, "and radar has seen UFO's."

Radar is no better than the radar observer and the radar observer has a mind. And where there's a mind there is the same old trouble. If the presentation on the radarscope doesn't look like it has looked for years a UFO is being tracked.

Radar is temperamental. The scope presentation of each radar has certain peculiarities and an operator gets used to seeing these. Occasionally, and for some unknown reason, these peculiarities suddenly change. For months a temperature inversion may cause 50 or 75 targets to appear on the radarscope. The operator has learned to recognize them and knows that they are caused by weather. They are not UFO's. But overnight something changes and now this same temperature inversion causes only one or two targets. The operator isn't used to seeing this and the targets are now UFO's.

Many times we'd stumble across the fact that after the first report of a UFO being tracked on radar the same identical type of target would be tracked again, many times. But by this time the operator would have learned that they were caused by weather and it wouldn't be reported to us.

It is interesting to note that, to my knowledge, there has never been a radar sighting classed as "unknown" when radarscope photos were taken. The reason is simple. The radar operator can take ample time to re-examine what he had to interpret in seconds during the actual sighting. Also, more experienced radar operators have a chance to examine the scope presentation.

Mixed in with the fact that there are few really qualified observers on this earth is the power of suggestion. About the time someone yells "UFO!" and points, all powers of reasoning come to a screeching halt.

We saw this happen day after day.

Few people I ever talked to, once they had decided they were looking at a UFO, stopped to calmly say to themselves, "Now couldn't this be a balloon, star, planet, or something else explainable?"

In one instance I traveled halfway across the United States to investigate a report made by a high ranking man in the State Department. An experienced observer. It was evening by the time I got to talk to him and after he'd excitedly told me all the pertinent facts, how this bright fight had "jumped across the sky," he said, "Want to see it? It's still there but it's not jumping now."

We went outside and there was Jupiter.

Then, there was the UFO over Dayton, Ohio, in the summer of 1952.

I first heard about it at home. It was about six in the evening when the phone rang and it was one of the tower operators at Patterson Field.

The tower operators at Lockbourne AFB in Columbus, Ohio, 60 miles east of Dayton, had spotted "three fiery spheres flying in a V- formation" over their base. Two F-84's had been scrambled to intercept and they were in the air right now. So far, the tower operator told me, the intercept had been unsuccessful because the objects were traveling "two to three thousand miles an hour" and were too high for the old F-84's.

He was monitoring the two jets' radio conversation and he put his telephone near the speaker.

I heard:

"At 28,000 and still above us."

"High speed."

"Headed toward Wright-Patterson."

"Low on fuel, going home."

I made it to my car in record time and took off toward Wright-Patterson, about twelve miles from where I was living.

It was still light, although the sun was low, and as I drove I kept looking toward the east. Nothing. I reached the gate, showed my pass to the guard, and had just written the whole thing off as another UFO report when I saw them.

They convinced me.

Off to the east of the airbase were three objects that can best be described as three half-sized suns.

By the time I arrived at base operations there were three or four dozen people on the ramp, all looking up.

The standard comment was: "Look at them go."

About this time a C-54 transport taxied up and stopped. It was the "Kittyhawk Flight" from Washington and I knew several people who got off.

One passenger, an officer from ATIC, ran up to me and handed me a roll of film.

"Here's some pictures of them," he said breathlessly. "I never thought I'd see one."

The next passengers I recognized were two other officers, Ph.D. psychologists from the Aero Medical Laboratory. I knew them because they had visited Blue Book many times collecting data for a paper they were writing on UFO's.

The title of the paper was to be: *The Psychological Aspects of UFO Sightings.*

Almost climbing over each other in their effort to tell their story they told me how they had watched the UFO's from the C-54. Both had seen them "dogfighting" between themselves.

"How fast were they going?" I asked.

"Like hell," was their only answer but the way they said it and the looks on their faces emphasized their statement.

The crowd on the ramp had increased by now and some of the newcomers had binoculars. The men with the binoculars were the focal point of several individual groups as they watched and gave blow-by-blow accounts.

Some of the crowd were talking about jet fighters and it suddenly dawned on me that just across the parking lot was the operations office of the local ADC jet outfit, the 97th Fighter Interceptor Squadron.

I ran over to interceptor operations and went in. I knew the duty officer because several times before the 97th people had chased balloons over Dayton. When I told him about the UFO's all I received was a rather uninterested stare. When I said they were over the base he did me the courtesy of going out to look.

He came running back in and hit the scramble button. Three minutes later two F-86's were headed UFOward. They soon disappeared but their vapor trails kept the tense crowd informed of their progress.

And believe me there was tension.

As the vapor trails spiraled up, first as two distinct plumes, and later only one—as they blended at altitude—more than one pilot standing on the ramp expressed his thankfulness for his unenviable position—on the ground watching.

The vapor trails thinned out and disappeared right under the three UFO's and it was obvious that the two jets had closed in.

Here were three that didn't escape.

That night the 97th Fighter Interceptor Squadron added three more balloons to their record. The F-86's had been able to climb higher than the F-84's.

The next morning photos confirmed the balloons. They had been tethered together and carried an instrument package.

I had been fooled. Two Ph.D psychologists who had studied UFO's had been fooled. A C-54 load of "experienced observers" (many pilots) had been fooled. The tower operators had been fooled and so had a hundred others.

This was an interesting sighting and we used to discuss it a lot. All of the observers later agreed that what made them so excited was the tower operator's announcement: "F-84's from Lockbourne are chasing three high speed objects." This set the stage and from then on no one even considered the fact that if the objects had been traveling 2000 or 3000 miles an hour they would have been long gone in the fifteen minutes we watched them.

Secondly, I found out that the C-54, a slow airplane, had actually overtaken and passed the balloons between Columbus and Dayton but none of the passengers I talked to had stopped to think of this.

And I'm positive that in our minds the balloons, which were about 40 feet in diameter and at 40,000 feet, looked a lot larger than they actually were.

I know the power of suggestion plays an important role in UFO sightings. Once you're convinced you're looking at a UFO you can see a lot of things.

But then there's the "unknowns."

Any good saucer fan—wild eyed or sober—will magnanimously concede that a certain percentage of the UFO sightings are the misidentification of known objects. They drag out the "unknowns" as the "proof."

Technically speaking, an "unknown" report is one that has been made by a reliable observer (not necessarily experienced). The report has been exhaustively investigated and analyzed and there is no logical explanation.

To this, the Air Force says: "The Air Force emphasizes the belief that if more immediate detailed objective observational data could have been obtained on the 'unknowns' these too could have been satisfactorily explained."

I think the Case of the Lubbock Lights is an excellent example of this. It is probably one of the most thoroughly investigated reports in the UFO files and it contained the most precise observational data we ever received. Scientists from far and near tried to solve it. It remained an "unknown."

The men who made the original sightings stuck by the case and furnished the "more detailed objective observational data" the Air Force speaks of.

The mysterious fights appeared again and instead of looking for something high in the air they looked for something low and found the solution.

The world famous Lubbock Lights were night flying moths reflecting the bluish-green light of a nearby row of mercury vapor street lights.

I will go a step further than the Air Force, however, and quote from a letter from ex-Lieutenant Andy Flues, once an investigator for Project Blue Book. Flues' statement sums up my beliefs and, I'm quite sure, the beliefs of everyone who has ever worked on Projects Sign, Grudge or Blue Book.

Flues wrote: "Even taking into consideration the highly qualified backgrounds of some of the people who made sightings, there was not one single case which, upon the closest analysis, could not be logically explained in terms of some common object or phenomenon."

The only reason there are any "unknowns" in the UFO files is that an effort is made to be scientific in making evaluations. And being scientific doesn't allow for any educated assuming of missing data or the passing of judgment on the character of the observer. However, this is closely akin to being forced to follow the Marquis of Queensbury rules in a fight with a hood. The investigation of any UFO sighting is an inexact science at the very best. Any UFO investigator, after a few months of being steeped in UFO lore and allowed a few scientific rabbit punches, can make the best of the "unknowns" look like a piece of well-holed Swiss cheese.

But regardless of what I say, or what the Air Force says, or what anyone says, we are stuck with flying saucers. And as long as people report unidentified objects in the air, it's the Air Force's responsibility to explain them.

Project Blue Book will live on.

No responsible scientist will argue with the fact that other solar systems may be inhabited and that some day we may meet those people. But it hasn't happened yet and until that day comes we're stuck with our Space Age Myth—the UFO.

Printed in the USA
CPSIA information can be obtained
at www.ICGtesting.com
LVHW041516260823
756140LV00001B/141